Optimization of Logistics
Theory and Practice

TURKAY YILDIZ

Copyright © 2014 Turkay Yildiz

All rights reserved.

ISBN-13: 978-1-329-88243-0

DEDICATION

To my parents...

TABLE OF CONTENTS

PREFACE x

Part I 12

Chapter 1. The Optimization Concept 12

    1.1 Introduction 13

    1.2 Types of optimization problems 14

    1.3 Optimization Methods 18

    1.4 Complexity of algorithms and problems 19

    References 22

Chapter 2. Typical Base Optimization Problems 24

    2.1 Assignment Problem 24

    2.2 Knapsack Problem 30

    2.3 Facility location problem 33

    2.4 Lot sizing problem 36

    2.5 Vehicle routing problems 40

    2.6 Scheduling problem 45

        Quay Crane/Yard Crane Scheduling 46

        Scheduling (Employees, Stevedore, etc.) 47

    2.7 Shortest Path Problems 50

        2.7.1 Snapshot of an example solved by using Dijkstra's algorithm for the shortest path problem 53

2.8 Travelling Salesman Problem 54

References 63

Chapter 3. Algorithms 86

3.1 Ant colony optimization 86

3.2 The cross-entropy (CE) method 90

3.3 The Dijkstra and the Bellman–Ford algorithm 91

3.4 The genetic algorithm (GA) 93

3.5 The Hungarian algorithm & the Jonker-Volgenant algorithm . 96

3.6 The particle swarm optimization 98

3.6.1 Snapshots of examples solved by using PSO 102

The Himmelblau test function 102

The "pen holder" test function 104

3.7 The simulated annealing method 106

3.7.1 Snapshots of some examples solved by using simulated annealing method 108

The "test tube holder" sample test function 108

The Levi13 sample test function 109

3.8 The Wagner-Whitin algorithm 111

References 112

Part II 126

Chapter 4. Reducing the Kullback-Leibler Distance: The Cross-Entropy Method for Optimization 126

1. Introduction 126

2. Literature review 127

3. The CE Method 130

3.1 The continuous CE optimization 133

3.2 The combinatorial CE optimization 135

4. Examples 138

4.1 Finding the minimum value of the Giunta test function 139

4.2 Finding the minimum value of the Ackley test function 141

4.3 Finding the minimum value of the Rastrigin test function 144

4.4 Finding the minimum value of the Hölder table test function 147

5. Conclusion 149

6. References 149

Chapter 5. Logistics systems and optimization strategies under uncertain operational environment 156

1. Introduction 156

2. Literature review 158

3. A Background: Mathematical programming 164

3.1 Stochastic linear programming with recourse 165

    Probabilistic/Chance-constrained programming: 166

    Stochastic non-linear programming with recourse: 166

3.2 A sample model with an uncertainty component 167

3.3 Initialization with data and simulation 168

4. Conclusion 173

5. References 174

Chapter 6. Transportation Network Design by Heuristic Methods. 180

1. Introduction 180

2. Literature review 181

3. Data and methods 184

3.1 Various scenarios and network configurations – Optimal solutions by heuristics methods 188

    Scenario 1 190

    Scenario 2 192

    Scenario 3 194

    Scenario 4 196

4. Research findings and discussions 198

5. Conclusion 198

6. References 199

ABOUT THE AUTHOR 208

PREFACE

This book comes out from the materials I used to refer while doing my research on the optimization issues in logistics. I brought together some of these materials to form a guidance material on the fundamentals of the optimization concepts along with my own studies on the application of optimization methods.

This book consists of two parts and six chapters. The first part of the book, which consists of three chapters, is about introduction to optimization with typical base problems and algorithms for solving problems. The second part of this book consists of three my own researches on the application of optimization methods.

Each chapter of this book is independent of each other. I hope you will find this book useful, informative, beneficial and appropriate for your needs.

Turkay Yildiz, M.A., Ph.D.

Part I

Chapter 1. The Optimization Concept

Optimization is already a part of our life. You might not be aware of it. You are already doing optimization in your life. For example, you wake up early in the morning and try to catch a bus or train on time. Your objective is to catch the bus/train on time while properly using your limited time early in the morning. You go to shopping and decide what to purchase within your budget constraint. Here, you try to maximize your satisfaction within your budget and in a time constraint. You choose the shortest and accessible paths to arrive from point A to B. You try to allocate more time for your leisure activities while considering allocating more time for studying and doing your home works.

There are so many examples in our lives that we try to optimize while maximizing/minimizing our objectives with many other constraints as such time, cost, etc.

Optimization also play important role in business organizations. Indeed, systematically applying optimization techniques in business processes can help organizations save millions of dollars by helping in reducing costs and eliminating redundant activities. In return, these savings might help in investing in areas that are more relevant and might make the organization more competitive.

Depending on the number of variables, optimization problems range from a very simple short definition of a problem to a more complex longer definitions. Retrieving the best solutions for these kind of problems highly depend on the capabilities of the algorithms and thus the computer power. Thanks to the recent advancement of computers with powerful CPUs and larger memories, many complex problems could be solved within seconds.

This chapter introduces the fundamentals of optimization. In subsections, types of optimization problems, optimization methods for solving optimization problems and the concept of complexity of algorithms and problems are introduced.

1.1 Introduction

Optimization is an old subject that has shown resurgence since the advent of computers and whose methods are applied in many fields: economics, management, planning, logistics, robotics, optimal design, engineering, signal processing, etc. (Allaire and Craig 2007). Optimization is a mature area because of the extensive research that has been conducted over the last 60 years (Arora 2007). The optimization is a broad topic that touches on the calculus of variations, operations research (optimization of management processes or decision) and optimal control (Allaire and Craig 2007). Optimization is an important tool in the decision science and in the analysis of physical systems (Nocedal and Wright, 1999). Optimization is everywhere, from business to engineering design, planning your vacation to your daily routine (Yang 2008).

Optimization is very important and relevant to virtually every discipline (Rangaiah 2010). Professional organizations must maximize profit and minimize cost (Yang 2008). The technical design is to maximize the performance of any product designed while minimizing costs at the same time (Yang 2008). The study of logistics systems typically involves several phases (Yalaoui *et al.* 2013):

- The objective of the first phase is the acquisition of information, which is to identify the parameters that affect the system. This could be profit, time, potential energy, or the quantity or combination of quantities that can be represented by a single number (Nocedal and Wright, 1999).
- The second, very important, phase is the development of a model to integrate all the parameters identified.
- The third phase focuses on the evaluation and analysis of the performance of the system using the original model. This

performance is understood by its quantitative and qualitative aspects. However, the model can be large and difficult to use. A phase of further simplification can then reduce the order of the system, while maintaining the characteristics and properties of the original model.
- Finally, an optimization phase is regularly used to improve system performance.

Any optimization problem has three basic ingredients (Arora 2007):

- The optimization variables, also called the design variables.
- Cost function, also called the objective function.
- The constraints expressed as equalities or inequalities.

The variables for the problem can be continuous or discrete. Depending on the types of variables and functions, we obtain a continuous variable, discrete variable, differentiable and non-differentiable problems (Arora 2007).

Transcript of an optimization problem in a mathematical formulation is a critical step in the process of solving the problem (Arora 2007). The decision consists of the following steps (Talbi 2009):
- Formulate the problem: In this first step, a decision problem is identified. Then, an initial problem statement is made.
- Model the problem: In this important step, is a mathematical model is constructed for the problem.
- Optimize the problem: Once the problem is modeled, the solving procedure generates a good solution to the problem. The solution may be optimal or suboptimal.
- Implement a solution: The solution is tested in practice by the decision maker and is implemented if it is acceptable.

1.2 Types of optimization problems
Many types of problems have been addressed and many different types of algorithms have been studied in the literature (Arora 2007). The

methodology has been used in various practical applications and the range of applications is constantly increasing (Arora 2007). Optimization problems can be divided into the following general categories depending on the type of decision variables, function (s) objective and constraints (Diwekar 2008):

- **Linear programming (LP):** The objective function and constraints are linear. The decision variables are involved scalar and continuous.
- **Nonlinear programming (NLP):** The function and / or objective constraints are not linear. The decision variables are scalar and continuous.
- **Integer Programming (IP):** The decision variables are scalar and integers.
- **Mixed integer linear programming (MILP):** The objective function and constraints are linear. The decision variables are scalar; some of them are integers, while others are continuous variables.
- **Mixed integer nonlinear programming (MINLP):** A problem of integer linear programming and continuous decision variables involved.
- **Discrete optimization:** problems involving discrete variables (integer) decision. This IP, MILP, and includes MINLPs.
- **Optimal Control:** The decision variables are vectors.
- **Stochastic programming or stochastic optimization:** It is also called optimization under uncertainty. In these problems, the objective function and / or constraints have uncertain variables (random). Often involves the above categories as subcategories.
- **Multi-objective optimization:** problems involving more than one objective. Often involves the above categories as subcategories.

The term "programming" in this context means planning activities that consume resources and / or meet the requirements as expressed in the constraints, not the other meaning – coding computer programs (Chen *et al.* 2011).

A linear programming problem (LP) is a class of mathematical programming problem, a problem of constrained optimization, in which we seek to find a set of values for continuous variables that maximizes or minimizes a linear objective function z, while satisfying a set of linear constraints (a system of simultaneous linear equations and / or inequalities) (Chen *et al.* 2011).

Linear programming is a powerful mathematical modeling technique, which is widely used, in business planning, technical design, the oil industry, and many other optimization applications (Yang 2008). Linear programming (LP) problems involve linear objective function and linear constraints (Diwekar 2008). The basic idea of linear programming is to find the maximum or minimum under linear constraints (Yang 2008).

An integer problem (linear) programming (IP) is a linear programming problem, in which at least one of the variables is restricted to integer values (Chen *et al.* 2011).

In the last two decades there has been an increasing use of a problem other term – **mixed integer programming (MIP)** – for LPs with integer restrictions on part or all of the variables (Chen *et al.* 2011).

A combinatorial optimization problem (COP) is a discrete optimization problem in which we seek to find a solution within a finite set of solutions that maximizes or minimizes an objective function (Chen *et al.* 2011). This type of problem usually arises in the selection of a finite set of mutually exclusive alternatives (Chen *et al.* 2011).

In Nonlinear (NLP) programming problems, either the objective function, constraints, or both the objective and constraints are non-linear (Diwekar 2008).

Discrete optimization problems can be classified as integer programming (IP) problems, the mixed linear integer programming (MILP), and non-linear programming (MINLP) mixed integer problems (Diwekar 2008).

Optimization under uncertainty: optimization under uncertainty refers to that branch of optimization where there are uncertainties in the data or model, and is popularly known as stochastic programming or stochastic optimization problems (Diwekar 2008). In this terminology, related to stochastic randomness and programming refers to mathematical programming techniques such as LP, NLP, IP, MILP, and MINLP (Diwekar 2008). The simplest approach to deal with uncertainty is to estimate the average value of each parameter and solve a deterministic problem (Talbi 2009).

Multi-objective optimization: Suppose we try to improve product performance while trying to minimize the cost at the same time (Yang 2014). In this case, we are dealing with a multi-objective optimization problem (Yang 2014). Many new concepts are needed to solve the multi-objective optimization (Yang 2014).

Dynamic optimization: In dynamic optimization problems, the objective function is deterministic at some point, but varies over time (Talbi 2009).

The main issues to solve dynamic optimization problems are Talbi (2009):

- Detect the change in the environment where it occurs. For most problems of real life, change is smooth rather than radical.
- Respond to the changing environment to follow the new global optimal solution. The objective is to dynamically change track the optimal solution as close as possible.

Robust Optimization: To solve the problem, we must consider the fact that a solution must be acceptable with respect to slight variations in the values of the decision variables (Talbi 2009). The term refers to the robust solutions (Talbi 2009). Robust optimization can be regarded as a specific type of problem with uncertainties (Talbi 2009).

In this class of problems, the solution implementation should be insensitive to small variations in the design parameters (Talbi 2009). This variation can be caused by production tolerances or drift parameters during operation (Talbi 2009).

1.3 Optimization Methods

There are three major classes of algorithms to solve problems of integer programming (Pop 2012):

- **Exact algorithms** are guaranteed to find the optimal solution, but it can take an exponential number of iterations. In practice, they are usually only applicable to small instances, due to long operating times caused by the complexity. They include: branch-and-bound, branch-and-cut, cutting plane, and dynamic programming algorithms.
- **Approximation algorithms** provide a sub-optimal solution in polynomial time, with a bounding constraint on the degree of sub-optimality. Unlike heuristics, which generally are pretty good solutions within a reasonable time, approximation algorithms provide provable solution quality and provable execution limits (Talbi 2009).
- **Heuristic algorithms** provide an optimum solution, but make no guarantee as to its quality. Heuristic algorithms use trial and error, learning and adapting to solve problems (Yang 2010). Although the running time is not guaranteed to be polynomial, empirical evidence suggests that some of these algorithms quickly find a good solution. These algorithms are particularly suited for large instances of optimization problems. Unlike exact methods, meta-heuristics are used to tackle cases of large-scale problems by providing satisfactory solutions within a reasonable time (Talbi 2009). Metaheuristics received increasing popularity over the last 20 years. Their use in many applications shows their effectiveness in solving large

complex problems (Talbi 2009). Modern metaheuristics algorithms are almost guaranteed to work well for a wide range of optimization problems difficult (Yang 2010).

Evolutionary computation techniques have attracted increasing attention in recent years to solve complex optimization problems (Sarker *et al.* 2002). They are more robust than the traditional methods based on formal logic or mathematical programming for many real world problems (Sarker *et al.* 2002). Evolutionary computation techniques can cope with complex problems of better optimization of conventional optimization techniques (Sarker *et al.* 2002).

Finally, it should be noted that there is "no free lunch" in the optimization (Yang 2010). It has been proved by Wolpert and Macready in 1997 that if the algorithm A is better than algorithm B for some problems, then B A to outperform other problems (Yang 2010). That is to say, a universally efficient algorithm does not exist (Yang 2010).

1.4 Complexity of algorithms and problems

The first step in the study of a combinatorial problem is whether the problem is "easy" or "hard" (Pop 2012). This classification is a task of complexity theory.

The efficiency of the algorithm is often measured by the computational complexity or computational complexity (Yang 2008). In the literature, this complexity is also called Kolmogorov complexity (Yang 2008). For a given problem size n, the complexity is noted using the Big-O notation as $O(n^2)$ or $O(n \log n)$ (Yang 2008).

$O(n)$ denotes a linear complexity while $O(n^2)$ has a quadratic complexity (Yang 2008). That is, if n is doubled, the time will double the linear complexity, but will quadruple the quadratic complexity (Yang 2008).

Complexity of algorithms (Talbi 2009): An algorithm needs two important resources to solve a problem: time and space. The time complexity of the algorithm is the number of steps required to solve a problem of size n. Complexity is usually defined as the worst-case analysis. The purpose of determining the computational complexity of an algorithm is not to obtain a number of correct but not an asymptotic bound on the number of steps. Big-O notation makes use of the asymptotic analysis. It is one of the most popular notations in the analysis of algorithms.

Big O notation uses a function to describe how the worst performance of the algorithm is the size of the problem size becomes very large (Stephens 2013). (This is sometimes called the asymptotic performance of the program) (Stephens 2013). The function is written in brackets after an uppercase O (Stephens 2013).

Complexity of the problems (Talbi 2009): The complexity of a problem is equivalent to the complexity of the best algorithm solving this problem. One problem is treatable (or easy) it is a polynomial time algorithm to solve it. One problem is insoluble (or difficult) if no polynomial time algorithm exists to solve the problem. The theory of the complexity of issues deals with decision problems. A decision problem always has a yes or no.

An important aspect of the theory of computation is to classify problems in complexity classes (Talbi 2009). **A complexity class** is the set of all problems that can be solved using a given amount of computing resources. There are two important classes of problems: P and NP (Talbi 2009).

Therefore, the P represents the family problems where a known polynomial-time algorithm exists to solve the problem. Problems in class P are then relatively easy to solve (Talbi 2009).

A problem is called non-deterministic polynomial (NP) if its solution can be guessed and evaluated in polynomial time, and there is no known rule to make such a proposal (hence, non-deterministic) (Yang

2008).

Some problems of the class P: Some classical problems in class P are minimum spanning tree, shortest path problems, network maximum flow, maximum matching bipartite, and programming linear continuous models (Talbi 2009).

The question of whether P = NP is one of the most important open questions due to the large impact, the answer would have on the theory of computational complexity (Talbi 2009). Obviously, for every problem in P we solve a non-deterministic algorithm (Talbi 2009).

A decision problem A NP is NP-complete if all other problems of the class NP are polynomially reduced to the problem A (Talbi 2009). If a polynomial deterministic algorithm exists to solve an NP-complete problem, then all problems in class NP can be solved in polynomial time (Talbi 2009).

No known algorithms exist to solve NP-hard problems, and approximate solutions or heuristic solutions are possible (Yang 2008). Thus, heuristics and metaheuristics are very promising in obtaining approximate solutions or suboptimal nearly optimal solutions / (Yang 2008). Many popular academic problems are NP-hard them (Talbi 2009):

- Sequencing problems and planning such as flow shop, job-shop planning, or open shop scheduling.
- Assignment and location problems such as quadratic assignment problem (QAP), generalized assignment problem (GAP), the location facility, and p -median problem.
- Consolidation of problems such as data clustering, graph partitioning and graph coloring.
- Routing and covering problems such as vehicle routing problem (VRP), set covering problem (SCP), the problem of Steiner tree and covering tour problem (CTP).

- Knapsack and packing/cutting problems, and so on.

References

Allaire, Gregoire; Craig, Alan (Translated by). Numerical Analysis and Optimization: An Introduction to Mathematical Modelling and Numerical Simulation. Cary, NC, USA: Oxford University Press 2007. p 294.

Arora, J. S. (2007). Optimization of Structural and Mechanical Systems. Hackensack, NJ, World Scientific.

Arora, Jasbir S. (Editor). Optimization of Structural and Mechanical Systems. River Edge, NJ, USA: World Scientific 2007. p 2.

Chen, Der-San; Batson, Robert G.; Dang, Yu. Applied Integer Programming Modeling and Solution. Hoboken, NJ, USA: Wiley 2011. p 3-7.

Nocedal, Jorge; Wright, Stephen J.. Numerical Optimization. Secaucus, NJ, USA: Springer, 1999. p 1-2.

Pop, Petrica C.; Versita (Contribution by). De Gruyter Series in Discrete Mathematics and Applications, Volume 1 : Network Design Problems : Modeling and Optimization of Generalized Network Design Problems. Hawthorne, NY, USA: Walter de Gruyter 2012. p 1-3.

Rangaiah, G. P. (2010). Stochastic Global Optimization : Techniques and Applications in Chemical Engineering. Singapore, World Scientific.

Sarker, Ruhul (Editor); Mohammadian, Masoud (Editor); Yao, Xin (Editor). Evolutionary Optimization. Secaucus, NJ, USA: Kluwer Academic Publishers 2002. p 10.

Stephens, Rod. Essential Algorithms : A Practical Approach to Computer Algorithms. Somerset, NJ, USA: Wiley 2013. p 31-32.

Talbi, El-Ghazali. (2009). Metaheuristics. Wiley.

Urmila Diwekar. (2008). Introduction to Applied Optimization. Springer US.

Yalaoui, Alice; Chehade, Hicham; Yalaoui, Farouk; Amodeo, Lionel. (2013). Optimization of Logistics. Wiley-ISTE.

Yang, Xin-She. (2014). Nature-Inspired Optimization Algorithms. Elsevier.

Yang, Xin-She. Engineering Optimization: An Introduction with Metaheuristic Applications. Hoboken, NJ, USA: Wiley 2010. p xxviii. Yang, Xin-She. Introduction to Mathematical Optimization: From Linear Programming to Metaheuristics. Cambridge, GBR: Cambridge International Science Publishing 2008. p 3-79.

Chapter 2. Typical Base Optimization Problems

This chapter briefly introduces typical pure base optimization problems of logistics systems. These problems are namely the assignment problem, knapsack problem, facility location problem, lot-sizing problem, vehicle routing problem, scheduling problem, shortest path problem, and travelling salesman problem. Typical optimization problems are not limited to these mentioned pure base problems in this chapter. Indeed, the real world problems usually require much more complicated problem definitions and utilization of sophisticated solvers for finding optimal solutions.

In each section of this chapter, mathematical representation of the optimization problem is shown. Additionally, the literature on this subject is so vast. Therefore, in each section the titles of the most recent and highly cited studies are shown in tables. Readers might take a look at these studies and further familiarize themselves for the recent studies on the optimization topic.

2.1 Assignment Problem

Assignment problem is a special case of transportation problem and the transportation problem is a special case of linear programming problem, so it is a special case of linear programming problem (Mishra and Agarwal 2009). The assignment problem can be considered as a special case of transportation problem (Williams 2013). It can be considered a problem with n sources and sinks n (Williams 2013). Each source has an availability of one unit and each well apply for 1 unit (Williams 2013). The objective of the assignment problem is to assign a number of tasks or jobs to an equal number of persons or the machine to a minimum or maximum cost profit (Mishra and Agarwal 2009). Assignment problem is a minimization problem (Mishra and Agarwal 2009).

The assignment problem is to find a maximum profit assignment of n tasks to n machines such that each task (i = 1, 2, ..., n) is assigned to exactly one machine (j = 1, 2, ..., n) and each machine is assigned to exactly one task. The general assignment problem (GAP) is a generalization of the assignment problem that finds a maximum profit assignment of m tasks to n (m > n) machines such that each task is assigned to exactly one machine and that each machine is allowed to be assigned to more than one task, subject to its capacity limitation (Chen et al. 2011).

The GAP is also known as the maximum weighted bipartite matching problem. Given a bipartite graph G(V,U,E) where V and U are two partitions and E edges between two partitions, the problem is the selection of a subset of the edges with maximum sum of weights such that each node $v \in V$ or $u \in U$ is connected to at most one edge.

The problem is finding a minimal cost (or maximal profit) assignment of n tasks over m capacity-constrained servers, whereby each task has to be processed by only one server.

Objective function

$$\text{Minimize } Z = \sum_{i=1}^{n}\sum_{j=1}^{m} c_{ij} x_{ij}$$

Subject to

$$\sum_{j=1}^{m} x_{ij} = 1, \qquad i = 1,...,n$$

$$\sum_{j=1}^{m} a_{ij} x_{ij} \leq b_j, \qquad j = 1,...,n$$

$$x_{ij} \in \{0,1\}, \qquad i = 1,...,n, \quad j = 1,...,m$$

where parameters are

n   = number of tasks

m = number of servers

C_{ij} = cost of assigning task i to server j

b_j = units of resource available to server j

a_{ij} = units of resource required to perform task i by server j

and variables

x_{ij} = 1 if task i is assigned to server j, 0 otherwise

GAP is to find a minimal cost (or maximum profit) assignment of n tasks on servers with limited capabilities m, each task should be handled by a single server (Sarker 2008). The general assignment problem has a wide range of relevant fields. For example, one such area is the situation causing in container terminals (Cheung *et al.* 2002; Zhang *et al.* 2002). Given an n tasks on servers with limited capacity m can be considered assignment of n number of straddle carriers (SC) or trailers with special containers to minimize the total time (t) consumed or cost (c) when handling the containers.

An assignment problem is a discrete optimization problem (Subbu and Sanderson 2004). Discrete optimization problems incur a heavy penalty because of dimensionality, because the size of the problems grows exponentially with the number of options along each dimension (Subbu and Sanderson 2004).

An extension of the complex assignment problem is the quadratic assignment problem (Williams 2013). This problem occurs when the "cost of an assignment" is not independent from other assignments. The resulting problem can be seen as a problem of assigning a quadratic objective function (Williams 2013).

Various algorithms exist for the solution of general assignment problem. To solve the problems of assignment in a container terminal, two algorithms, i.e., Jonker-Volgenant and Hungarian algorithms are considered. Both algorithms provide the same solutions and the solutions are considered to be the best possible optimal solutions for the given probability matrix. For a performance comparison of the both algorithms, see Figure 2.1.

The most recent and highly cited studies about the assignment problem are shown in Table 2.1.

Table 2.1 Studies in the literature about the assignment problem

A computational study of exact knapsack separation for the generalized assignment problem (Avella *et al.* 2010).
Solving the class responsibility assignment problem in object-oriented analysis with multi-objective genetic algorithms (Bowman *et al.* 2010).
The multi-unit assignment problem: theory and evidence from course allocation at harvard (Budish and Cantillon 2012).
The wiener maximum quadratic assignment problem (cela *et al.* 2011).
An efficient solution algorithm for solving multi-class reliability-based traffic assignment problem (Chen *et al.* 2011).
The storage location assignment problem for outbound containers in a maritime terminal (Chen and Lu 2012).
An ant colony optimisation algorithm for solving the asymmetric traffic assignment problem (D'acierno *et al.* 2012).
Exploiting group symmetry in semidefinite programming relaxations of the quadratic assignment problem (de Klerk and Sotirov 2010).
State covariance assignment problem (Khaloozadeh and Baromand 2010).
A multi-objective evolutionary algorithm for the deployment and power assignment problem in wireless sensor networks (Konstantinidis *et al.* 2010).
Reliability optimization of component assignment problem for a multistate network in terms of minimal cuts (Lin and Yeh 2011).
Stability of user-equilibrium route flow solutions for the traffic assignment problem (Lu and Nie 2010).
Grasp with path-relinking for the generalized quadratic assignment problem (Mateus *et al.* 2011).
Multiobjective genetic algorithms for solving the impairment-aware routing and wavelength assignment problem (Monoyios and Vlachos 2011).
A class of bush-based algorithms for the traffic assignment problem (Nie 2010).
A cell-based merchant-nemhauser model for the system optimum dynamic traffic assignment problem (Nie 2011).
Solving the dynamic user optimal assignment problem considering queue spillback (Nie and Zhang 2010).
Bees algorithm for generalized assignment problem (Ozbakir *et al.* 2010).
A two phase method for multi-objective integer programming and its application to the assignment problem with three objectives (Przybylski *et al.* 2010).
Partial eigenvalue assignment problem of high order control systems using orthogonality relations (Ramadan and El-Sayed 2010).
A due-date assignment problem with learning effect and deteriorating jobs (Wang and Guo 2010).
Single-machine due-window assignment problem with learning effect and deteriorating jobs (Wang and Wang 2011).
Effective formulation reductions for the quadratic assignment problem (Zhang *et al.* 2010).
Routing and wavelength assignment problem in pce-based wavelength-switched optical networks (Zhao *et al.* 2010).
A novel global harmony search algorithm for task assignment problem (Zou *et al.* 2010).

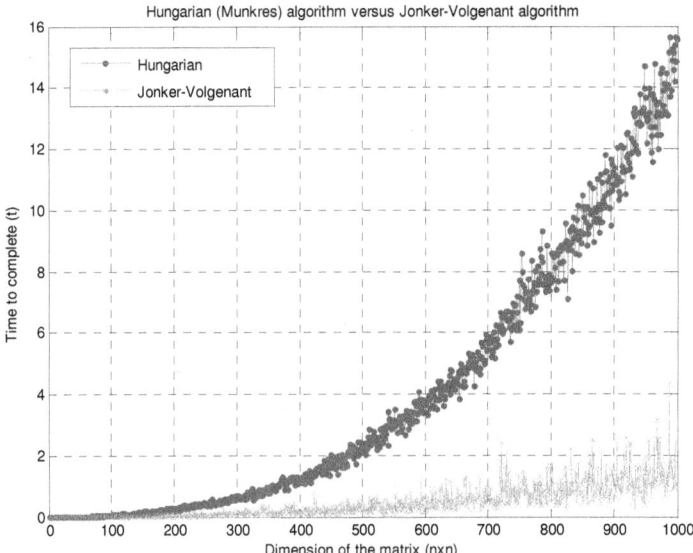

Figure 2.1 Hungarian (Munkres) algorithm versus Jonker-Volgenant algorithm

2.2 Knapsack Problem

The "knapsack" stems from the rather artificial application to try to fulfill a hiker's knapsack at the total maximum value (Williams 2013) application. Each item it considers taking with it has a certain value and a certain weight (Williams 2013). Limiting overall weight is the single constraint (Williams 2013).

The knapsack problem the has received considerable attention in the literature during the early development of operations research algorithms (1950-1970), mainly because it can be used as a sub problem in the development of an decomposition algorithm to the well-known stock cut (or trim loss) problem and because of a general linear integer problem can be converted into a knapsack problem (Chen *et al.* 2011). Dozens of specialized algorithms for knapsack problems have been developed, which includes dynamic programming, enumeration, Lagrange multiplier, and the network approaches (Chen *et al.* 2011).

The knapsack problem takes a set of items, each with a weight and profit, and tries to fit as many of these items in the knapsack to maximize profit while not exceeding the maximum weight the knapsack may contain (Rondeau and Bostian 2009).

The name is derived from a problem of decision facing a hiker who is to select a set of given elements to be included in his backpack (or knapsack) within a specified weight (Chen *et al.* 2011). Each selected element contributes a value (relative) for hiking and the purpose of this decision problem is to maximize the total value of all the selected items (Chen *et al.* 2011).

The problem of the knapsack is sometimes called the problem of loading cargo when shipments of different weights are selected for loading on a vessel with a weight capacity limited (Chen *et al.* 2011).

Objective function

$$\text{Maximize } Z = \sum_{i=1}^{m}\sum_{j=1}^{n} v_j x_{ij}$$

Subject to

$$\sum_{i=1}^{m} x_{ij} \leq 1, \quad \forall j$$

$$\sum_{j=1}^{n} w_j x_{ij} \leq C_i, \quad \forall j$$

$$x_{ij} \in \{0,1\}, \quad \forall i, j$$

where parameters are

m = number of containers (index i)

n = number of items (index j)

w_j = weight of item j

v_j = value of item j

C_i = capacity of container i (weight)

The most recent and highly cited studies about the knapsack problem are shown in Table 2.2.

Table 2.2 Studies in the literature about the knapsack problem

Kernel search: a general heuristic for the multi-dimensional knapsack problem (Angelelli *et al.* 2010).
Identifying preferred solutions to multi-objective binary optimisation problems, with an application to the multi-objective knapsack problem (Argyris *et al.* 2011).
A multi-level search strategy for the 0-1 multidimensional knapsack problem (Boussier *et al.* 2010).
A knapsack problem as a tool to solve the production planning problem in small foundries (Camargo *et al.* 2012).
A column generation method for the multiple-choice multi-dimensional knapsack problem (Cherfi and Hifi 2010).
A heuristic approach for allocation of data to rfid tags: a data allocation knapsack problem (dakp) (Davis *et al.* 2012).
Revenue maximization in the dynamic knapsack problem (Dizdar *et al.* 2011).
Development of core to solve the multidimensional multiple-choice knapsack problem (Ghasemi and Razzazi 2011).
Sensor selection in distributed multiple-radar architectures for localization: a knapsack problem formulation (Godrich *et al.* 2012).
A ptas for the chance-constrained knapsack problem with random item sizes (Goyal and Ravi 2010).
Improved convergent heuristics for the 0-1 multidimensional knapsack problem (Hanafi and Wilbaut 2011).
Problem reduction heuristic for the 0-1 multidimensional knapsack problem (Hill *et al.* 2012).
An ant colony optimization approach for the multidimensional knapsack problem (Ke *et al.* 2010).
Fully polynomial approximation schemes for a symmetric quadratic knapsack problem and its scheduling applications (Kellerer and Strusevich 2010).
Upper bounds for the 0-1 stochastic knapsack problem and a b&b algorithm (Kosuch and Lisser 2010).
Assessing solution quality of biobjective 0-1 knapsack problem using evolutionary and heuristic algorithms (Kumar and Singh 2010).
Reoptimization in lagrangian methods for the 0-1 quadratic knapsack problem (Letocart *et al.* 2012).
Interdicting nuclear material on cargo containers using knapsack problem models (Mclay *et al.* 2011).
The multidimensional knapsack problem: structure and algorithms (Puchinger *et al.* 2010).
Fusing ant colony optimization with lagrangian relaxation for the multiple-choice multidimensional knapsack problem (Ren *et al.* 2012).
Dynamic programming based algorithms for the discounted {0-1} knapsack problem (Rong *et al.* 2012).
A two state reduction based dynamic programming algorithm for the bi-objective 0-1 knapsack problem (Rong *et al.* 2011).
A cooperative local search-based algorithm for the multiple-scenario max-min knapsack problem (Sbihi 2010).
An effective hybrid eda-based algorithm for solving multidimensional knapsack problem (Wang 2012).
Solving 0-1 knapsack problem by a novel global harmony search algorithm (Zou *et al.* 2011).

2.3 Facility location problem

Facility location problems deal with the number, location, equipment and size of new plants as well as the sale, removal or reduction of existing facilities (Ghiani 2013). In the logistics business, the process of site planning is to design the entire facility through which revenue from suppliers to demand points, while in the public sector, it is to determine all facilities from which users are served (Ghiani 2013).

One of the most important aspects of logistics is deciding where to locate new facilities such as retailers, warehouses or factories (Bramel and Simchi-Levi, 1997). These strategic decisions are a key determinant of whether the materials circulate efficiently in the distribution system (Bramel and Simchi-Levi, 1997). When location decisions are needed: the location of the installation decisions must obviously be taken when a logistics system is started from zero (Ghiani 2013). They are also required as a result of variations in the pattern of demand or the spatial distribution, or due to changes in materials, energy and labor costs.

Mathematical models of location are designed to address a number of issues, including (Daskin 2013)

(a) How many facilities should be located?
(b) Where each facility should be located?
(c) How large should each facility be?
(d) How should require the services of facilities allocated to installations?

In general, the goal is to find a set of facilities so that the total cost is minimized subject to a number of constraints that could include (Bramel and Simchi-Levi, 1997):

- each warehouse has a capacity that limits the area that can provide.
- Each retailer receives deliveries of one and only one warehouse.

- each retailer must be at a fixed distance from the warehouse that supplies it, so that a reasonable delivery lead time is guaranteed.

Facility location is a wide range of mathematical models, methods and applications in operations research (Chvátal 2011). This is an interesting topic for theoretical studies, experimental research and real-world applications (Chvátal 2011).

The most recent and highly cited studies about the facility location problem are shown in Table 2.3.

Table 2.3 Studies in the literature about the facility location problem

A computational comparison of several formulations for the multi-period incremental service facility location problem (Albareda-Sambola et al. 2010).
The facility location problem with bernoulli demands (Albareda-Sambola et al. 2011).
P-hub approach for the optimal park-and-ride facility location problem (Aros-Vera et al. 2013).
Semi-lagrangian relaxation applied to the uncapacitated facility location problem (Beltran-Royo et al. 2012).
Solving conflicting bi-objective facility location problem by nsga ii evolutionary algorithm (Bhattacharya and Bandyopadhyay 2010).
An optimal bifactor approximation algorithm for the metric uncapacitated facility location problem (Byrka and Aardal 2010).
A computational study of a nonlinear minsum facility location problem (Carrizosa et al. 2012).
Strategic closed-loop facility location problem with carbon market trading (Diabat et al. 2013).
A primal-dual approximation algorithm for the facility location problem with submodular penalties (Du et al. 2012).
An approximation algorithm for the k-level capacitated facility location problem (Du et al. 2010).
A new approximation algorithm for the multilevel facility location problem (Gabor and van Ommeren 2010).
A generalized weiszfeld method for the multi-facility location problem (Iyigun and Ben-Israel 2010).
The ordered capacitated facility location problem (Kalcsics et al. 2010).
Competitive facility location problem with attractiveness adjustment of the follower a bilevel programming model and its solution (Kucukaydin et al. 2011).
Integrated use of fuzzy c-means and convex programming for capacitated multi-facility location problem (Kucukdeniz et al. 2012).
An efficient genetic algorithm for solving the multi-level uncapacitated facility location problem (Maric 2010).
The discrete facility location problem with balanced allocation of customers (Marin 2011).
An integer decomposition algorithm for solving a two-stage facility location problem with second-stage activation costs (Penuel et al. 2010).
On the structure of the solution set for the single facility location problem with average distances (Puerto and Rodriguez-Chia 2011).
The reliable facility location problem: formulations, heuristics, and approximation algorithms (Shen et al. 2011).
A tabu search heuristic procedure for the capacitated facility location problem (Sun 2012).
An improved benders decomposition algorithm for the logistics facility location problem with capacity expansions (Tang et al. 2013).
A genetic algorithm for the uncapacitated facility location problem (Tohyama et al. 2011).
An approximation algorithm for the k-level stochastic facility location problem (Wang et al. 2010).
A cut-and-solve based algorithm for the single-source capacitated facility location problem (Yang et al. 2012).

2.4 Lot sizing problem

Once the levels of inventory are determined, the next step is to calculate in what quantities the inventory will be replaced. This is called lot sizing. The lot size is the amount of material to be ordered from a supplier or produced internally to meet demand.

There are nine major types of lot-size methods, which fit into the following two categories:

Demand-based methods (static): Order quantities are kept constant.

- Fixed order quantity: min/max
- Economic order quantity (EOQ): is calculated periodically and used as fixed order quantity during interim

Discrete method (dynamic): Order quantities vary.

- Period order quantity
- Lot-for-lot
- Periods of supply
- Least unit cost
- Least total cost
- Part-period balancing
- Wagner-Whitin algorithm

Selecting the appropriate combination of methods will help reduce the ordering, setup and transportation costs, and reduce overall inventory levels of products being manufactured (Viale and Carrigan, 1996):

- **Fixed Order Quantity:** method of fixed order quantity orders will always suggest planned to be released for a predetermined fixed amount. The predetermined amount can be determined on the basis of experience and / or the use of the technique of the economic order quantity.
- **The economic order quantity (EOQ)** is the other type of formula based on demand or static. This calculation determines the amount to be purchased or made by determining the minimum cost of purchase or construction with the cost to

carry inventory. The formula can be used to determine the minimum units to be built or purchased, or the minimum cost in dollars.

- **Period Order Quantity** Period order quantity is a lot-sizing technique where the lot size is equal to the requirements of a given number of periods in the future. The period order quantity is similar to the period of supply, with the exception of the order cycle is based on the calculation EOQ. The order frequency and the order quantities are scheduled using this method.
- **Lot-for-Lot** This is a technique commonly used in MRP lot-sizing technique in just-in-time (JIT) situations, in collaboration with the safety stock. In this method, the planned orders are generated at the height of the net requirements in each period.
- **Periods of supply** This method establishes- mainly by experience-an amount of the order to cover a predetermined period of time.
- **Least unit cost** The least unit cost method adds the cost of ordering and inventory-carrying cost for each trial lot size and divides the number of units in the lot size. Lot size with the lowest unit cost is chosen.
- **Least total cost** The least total cost lot-sizing technique calculates the order quantity by comparing the set (or order) costs and carrying costs for different lot sizes and selects the lot size where these costs are nearest equal.
- **Part-Period Balancing** This technique is similar to the method in the least cost method. However, this method uses a routine called look forward / looking back. When the function look forward / looking back is used, a lot quantity is calculated, and before it is firmed up, the next or previous period is examined to determine whether it would be economical to fit in the current lot.
- **Wagner-Whitin Algorithm** The last method is the algorithm of Wagner-Whitin. This is a very complex process that evaluates

all possible means to meet the needs of each period of the planning horizon.

The most recent and highly cited studies about the lot-sizing problem are shown in Table 2.4.

Table 2.4 Studies in the literature about the lot sizing problem

Uncapacitated lot-sizing problem with production time windows, early productions, backlogs and lost sales (Absi et al. 2011).
Adaptive genetic algorithm for lot-sizing problem with self-adjustment operation rate: a discussion (Cardenas-Barron 2010).
A volume flexible economic production lot-sizing problem with imperfect quality and random machine failure in fuzzy-stochastic environment (Das et al. 2011).
A particle swarm optimization for solving joint pricing and lot-sizing problem with fluctuating demand and unit purchasing cost (Dye and Hsieh 2010).
A particle swarm optimization for solving joint pricing and lot-sizing problem with fluctuating demand and trade credit financing (Dye and Ouyang 2011).
Solving single-product economic lot-sizing problem with non-increasing setup cost, constant capacity and convex inventory cost in o(n log n) time (Feng et al. 2011).
A heuristic approach for a multi-product capacitated lot-sizing problem with pricing (Gonzalez-Ramirez et al. 2011).
Stochastic lot-sizing problem with inventory-bounds and constant order-capacities (Guan and Liu 2010).
A robust lot sizing problem with ill-known demands (Guillaume et al. 2012).
A simulation metamodelling based neural networks for lot-sizing problem in mto sector (Hachicha 2011).
A fix-and-optimize approach for the multi-level capacitated lot sizing problem (Helber and Sahling 2010).
A robust block-chain based tabu search algorithm for the dynamic lot sizing problem with product returns and remanufacturing (Li et al. 2014).
Integrating run-based preventive maintenance into the capacitated lot sizing problem with reliability constraint (Lu et al. 2013).
A new algorithmic approach for capacitated lot-sizing problem in flow shops with sequence-dependent setups (Mohammadi et al. 2010).
Grasp heuristic with path-relinking for the multi-plant capacitated lot sizing problem. (Nascimento et al. 2010).
A simple fptas for a single-item capacitated economic lot-sizing problem with a monotone cost structure (Ng et al. 2010).
An o(t-3) algorithm for the capacitated lot sizing problem with minimum order quantities (Okhrin and Richter 2011).
The economic lot-sizing problem with remanufacturing and one-way substitution (Pineyro and Viera 2010).
Solving the stochastic dynamic lot-sizing problem through nature-inspired heuristics (Piperagkas et al. 2012).
A new silver-meal based heuristic for the single-item dynamic lot sizing problem with returns and remanufacturing (Schulz 2011).
An efficient computational method for a stochastic dynamic lot-sizing problem under service-level constraints (Tarim et al. 2011).
Stochastic dynamic lot-sizing problem using bi-level programming base on artificial intelligence techniques (Wong et al. 2012).
An hnp-mp approach for the capacitated multi-item lot sizing problem with setup times (Wu et al. 2010).
An mip-based interval heuristic for the capacitated multi-level lot-sizing problem with setup times (Wu et al. 2012).
A lagrangian relaxation based approach for the capacitated lot sizing problem in closed-loop supply chain (Zhang et al. 2012).

2.5 Vehicle routing problems

Given the fluctuations and the upward trend in the price of oil, transportation costs represent a share of more and more of the final cost charged to the customer (about 10-20% of the overall cost of doing business) it becomes essential to control these costs within global supply chains (Jarboui *et al.* 2013). The class of problems that result from these studies is commonly called the vehicle routing problem (VRP) (Jarboui *et al.* 2013). The classic problem of the development of vehicle routing is to construct roads with minimal cost to a set of vehicles can visit a set of customers geographically distributed exactly once (Jarboui *et al.* 2013).

The problem is to ascertain the operation plan satisfying the demand at various zones at minimum cost (Bish *et al.* 2001; Kim and Bae, 1998; Vis and De Koster 2003).

Objective function is

$$\text{Minimize } f_{obj} = \sum_{i=1}^{G}\sum_{j=1}^{Z}\sum_{k=1}^{F} C_{ijk} x_{ijk}$$

Subject to

$$\sum_{i=1}^{G}\sum_{k=1}^{F} L_k x_{ijk} \geq D_j \qquad \forall j$$

$$\sum_{j=1}^{Z}\sum_{k=1}^{F} L_k x_{ijk} \leq S_j \qquad \forall i$$

$$\sum_{j=1}^{Z} L_k x_{ijk} \leq U_{ki} \qquad \forall k, i$$

$$x_{ijk} \geq 0 \qquad \forall i, j, k$$

where parameters are

G = Number of source locations (index i)

Z = Number of receiving nodes for containers (index j)

F = Number of trailers available (index k)

L_k = Load capacity of trailer k

S_i = Quantity of available containers for transportation from location i

D_j = Quantity of containers required by zone j

C_{ijk} = Unit cost of transporting from location i to zone j by trailer k

U_{ik} = Maximum allowable containers that can be transported from location i by trailer k in a given period

and variables

x_{ijk} = the number of trips required by trailer k from location i to zone j

In the area of freight transport, both activities are generally distinguished (Jarboui *et al.* 2013):

- Full-load Transportation, the activity of a vehicle consists of a series of movements between the origins and destinations of the goods transported. For each movement, the goods to one customer is on board the vehicle.
- Less-than-truckload (LTL) transportation, each vehicle serves a range of customers from a warehouse where the goods are loaded. One of the central problems in the transportation of LTL is the vehicle routing problem (VRP)

Freight activities are essential in planning logistics systems. The reason is twofold (Ghiani 2013)

- First, they determine the most important part (often between one third and two thirds) of logistics costs; and,
- Second, they significantly affect the level of service provided to customers (Ghiani 2013).

Providing efficient a cost effective transport freight services results in an increase of the distance to the facilities of logistics system can be implemented economically (Ghiani 2013).

Problems related to the determination of the optimal routes for vehicles of one or more depots for a set of locations / customers are known as the vehicle routing problem (VRP) (Pop 2012). They have many practical applications in the field of distribution and logistics (Pop 2012).

The vehicle routing problem is, in practice, the most important extension of the traveling salesman problem where you have to schedule a number of vehicles, limited capacity, around a number of clients (Williams 2013). Thus, in addition to the sequence of customers to visit (the traveling salesman problem) one need to decide which vehicles visiting which customers (Williams 2013). Each customer has a known demand (assumed to be a commodity, but it can easily be extended to more than one product) (Williams 2013). Therefore, the limited capacities of vehicles need to be taken into account.

The distribution problem or vehicle routing (VRP) is often described as the problem in which vehicles based on a central depot need to visit geographically dispersed customers to meet the known demands of customers (Pardalos *et al.* 2002). The problem is to build a low-cost, feasible set of routes - one for each vehicle (Pardalos *et al.* 2002). A route is a sequence of locations that the vehicle has to go along with an indication of the service it provides (Pardalos *et al.* 2002). The vehicle must begin and end his tour at the station (Pardalos *et al.* 2002). We can say that the problem is a generalization of the traveling salesman problem (Pardalos *et al.* 2002). The traveling salesman problem (TSP) requires the determination of a minimum cost cycle that passes through each node of a given once (Pardalos *et al.* 2002) graph.

Due to the simplicity of the VRP, variations of the VRP, built on the basic VRP with additional features, proved more attractive to many researchers (Pop 2012):

1. The Capacitated VRP, wherein each vehicle has a finite capacity and each location has a finite demand.
2. The VRP with time windows, in which there is a specific opportunity in which to visit each location time window.
3. VRP with multiple depots, which generalizes the idea of a depot, in such a way that there are several depots from which each customer can be served.
4. The multi-commodity VRP wherein each location has associated demand for various commodities and each vehicle has a set of compartments in a single product which can be loaded. The problem then becomes one of deciding which products to place in which compartments to minimize the distance traveled.
5. The general routing problem (GVRP) vehicles is the problem of the design of delivery or collection of optimal routes, one given to a number of nodes, predefined sets, mutually exclusive and exhaustive node-sets (clusters) subject to capacity restrictions. The GVRP can be seen as a particular kind of location-routing problem (see, e.g. Laporte, Nagy and Salhi) for which several heuristic algorithms, mostly exist.

The most recent and highly cited studies about the vehicle routing problem are shown in Table 2.5.

Table 2.5 Studies in the literature about the vehicle routing problem

An exact algorithm for a vehicle routing problem with time windows and multiple use of vehicles (Azi et al. 2010).
Some applications of the generalized vehicle routing problem (Baldacci et al. 2010).
New route relaxation and pricing strategies for the vehicle routing problem (Baldacci et al. 2011).
Metaheuristics for the waste collection vehicle routing problem with time windows, driver rest period and multiple disposal facilities (Benjamin and Beasley 2010).
A tabu search algorithm for the heterogeneous fixed fleet vehicle routing problem (Brandao 2011).
Iterated variable neighborhood descent algorithm for the capacitated vehicle routing problem (Chen et al. 2010).
Branch-and-price-and-cut for the split-delivery vehicle routing problem with time windows (Desaulniers 2010).
A multi-start evolutionary local search for the two-dimensional loading capacitated vehicle routing problem (Duhamel et al. 2011).
An iterative route construction and improvement algorithm for the vehicle routing problem with soft time windows (Figliozzi 2010).
An improved multi-objective evolutionary algorithm for the vehicle routing problem with time windows (Garcia-najera and Bullinaria 2011).
Using simulated annealing to minimize fuel consumption for the time-dependent vehicle routing problem (Kuo 2010).
Application of genetic algorithms to solve the multidepot vehicle routing problem (Lau et al. 2010).
An enhanced ant colony optimization (eaco) applied to capacitated vehicle routing problem (lee et al. 2010).
A hybrid genetic - particle swarm optimization algorithm for the vehicle routing problem (Marinakis and Marinaki 2010).
A hybrid particle swarm optimization algorithm for the vehicle routing problem (Marinakis et al. 2010).
A memetic algorithm for the multi-compartment vehicle routing problem with stochastic demands (Mendoza et al. 2010).
A penalty-based edge assembly memetic algorithm for the vehicle routing problem with time windows (Nagata et al. 2010).
An effective memetic algorithm for the cumulative capacitated vehicle routing problem (Ngueveu et al. 2010).
The two-echelon capacitated vehicle routing problem: models and math-based heuristics (Perboli et al. 2011).
A hybrid evolution strategy for the open vehicle routing problem (Repoussis et al. 2010).
An adaptive large neighborhood search heuristic for the cumulative capacitated vehicle routing problem (Ribeiro and Laporte 2012).
A parallel heuristic for the vehicle routing problem with simultaneous pickup and delivery (Subramanian et al. 2010).
An artificial bee colony algorithm for the capacitated vehicle routing problem (Szeto et al. 2011).
An ant colony optimization model: the period vehicle routing problem with time windows (Yu and Yang 2011).
A parallel improved ant colony optimization for multi-depot vehicle routing problem (Yu et al. 2011).

2.6 Scheduling problem

The main activity of an industrial company in the short term is to organize production on a time frame (Yalaoui *et al.* 2013). Scheduling of production orders is to decide which machine(s) to use and in what order to achieve the production (Yalaoui *et al.* 2013). Because of the diversity in production processes and management styles that vary across industries, many methods for optimizing the allocation of tasks have been developed (Yalaoui *et al.* 2013).

In a scheduling problem, four basic concepts are involved: tasks (jobs), resources, constraints and objectives (Yalaoui *et al.* 2013). A job is defined by a set of operations that must be performed (Yalaoui *et al.* 2013). It is characterized in time by a start date and end date. A resource is a machine or a human involved in performing the work (Yalaoui *et al.* 2013). Constraints represent limitations in time, technology and resources. The objectives are the criteria to be optimized in terms of time, resources, costs or output (Yalaoui *et al.* 2013).

The purpose of the supply chain scheduling is to optimize short-medium term decisions in supply chains, taking into account the balance between the tangible economic objectives such as cost reduction or profit maximization and less tangible goals such as customer satisfaction or level of customer service (Sawik 2011). In addition to production scheduling, scheduling for the supply chain considering the manufacture and supply of materials and distribution of finished products, and includes other decisions related to the functional, spatial, and intertemporal integration and coordination of schedules for these activities (Sawik 2011).

The scheduling problem is known to be intractable computation in many cases (El-Abd and Rewini-El-Barr 2005). Fast optimal algorithms can only be obtained when restrictions are imposed on models representing the program and the distributed system (El-Abd and Rewini-El-Barr 2005). Solving the general problem in a reasonable time requires the use of heuristic algorithms (El-Abd and Rewini-El-Barr 2005). These heuristics do not guarantee optimal solutions to the problem, but they are trying to find solutions near optimal (El-Abd and Rewini-El-Barr 2005).

Quay Crane/Yard Crane Scheduling

The scheduling problem with the assumption is that there are n jobs and m machines. Each job must be processed on all machines (i.e. cranes) in a given order. A machine (i.e. crane) can only process one job at a time, and once a job is started on any machine (i.e. crane), it must be processed to completion. The objective is to minimize the sum of the completion times of all the jobs.

Objective function

$$\text{Minimize Z} = \sum_{j=1}^{n} t_{j(m),j}$$

Subject to

$$t_{j(r+1),j} \geq t_{j(r),j} + P_{j(r),j} \quad \text{for } r = 1, 2, ..., m-1 \text{ and } \forall j$$

$$t_{ij} - t_{ik} \leq -P_{ij} + U(1 - x_{ijk}) \quad \forall i, j, k$$

$$t_{ik} - t_{ij} \leq -P_{ik} + U x_{ijk} \quad \forall i, j, k$$

$$t_{ij} \geq 0 \quad \forall i, j$$

$$x_{ijk} \in \{0, 1\} \quad \forall i, j, k$$

where parameters are

n   = the number of jobs

m   = the number of machines

P_{ij} = the processing time of job j on machine i

j(r)   = the order of machines/operations for job j (for example, job j must be processed on machine 2 first (r=1,i=2), and then machine 4 (r=2, i=4), and so on). For any job j, r = m means the last operation of the job.

and variables:

t_{ij} = the start time of job j on machine i

x_{ijk} = 1 if job j precedes job k on machine i, 0 otherwise (i.e., if job k precedes job j on machine i)

Scheduling (Employees, Stevedore, etc.)

The problem is to determine the number of employees required to meet the different daily work force necessities of seaport terminal while minimizing the general scheduling cost.

Objective function

$$\text{Minimize } Z = \sum_{i=1}^{N} C_i x_i$$

Subject to

$$\sum_{i \in M_j} x_i \geq R_j \quad \forall j$$

$$x_i \geq 0 \quad \forall i$$

where parameters are

N = the total number of roster type

M_j = the set of roster types that will allow working on a day j

R_j = the number of employees required on each day j

C_i = weekly cost per employee assigned to roster type i

and variables

x_i = the number of employees assigned to roster type i

The most recent and highly cited studies about the scheduling problem are shown in Table 2.6.

Table 2.6 Studies in the literature about the scheduling problem

A neurogenetic approach for the resource-constrained project scheduling problem (Agarwal *et al.* 2011).
An artificial immune algorithm for the flexible job-shop scheduling problem (Bagheri *et al.* 2010).
A two-agent single-machine scheduling problem with truncated sum-of-processing-times-based learning considerations (Cheng *et al.* 2011).
An improved genetic algorithm for the distributed and flexible job-shop scheduling problem (de Giovanni and Pezzella 2010).
New multi-objective method to solve reentrant hybrid flow shop scheduling problem (Dugardin *et al.* 2010).
A novel competitive co-evolutionary quantum genetic algorithm for stochastic job shop scheduling problem (Gu *et al.* 2010).
A proactive approach for simultaneous berth and quay crane scheduling problem with stochastic arrival and handling time (Han *et al.* 2010).
A survey of variants and extensions of the resource-constrained project scheduling problem (Hartmann and Briskorn 2010).
A two-machine flowshop scheduling problem with deteriorating jobs and blocking (Lee *et al.* 2010).
A single-machine scheduling problem with two-agent and deteriorating jobs (Lee *et al.* 2010).
A single-machine learning effect scheduling problem with release times (Lee *et al.* 2010).
A hybrid tabu search algorithm with an efficient neighborhood structure for the flexible job shop scheduling problem (Li *et al.* 2011).
A pareto approach to multi-objective flexible job-shop scheduling problem using particle swarm optimization and local search (Moslehi and Mahnam 2011).
A local-best harmony search algorithm with dynamic sub-harmony memories for lot-streaming flow shop scheduling problem (Pan *et al.* 2011).
A discrete artificial bee colony algorithm for the lot-streaming flow shop scheduling problem (Pan *et al.* 2011).
An iterated greedy algorithm for the flowshop scheduling problem with blocking (Ribas *et al.* 2011).
The hybrid flow shop scheduling problem (Ruiz and Vazquez-Rodriguez 2010).
Total flow time minimization in a flowshop sequence-dependent group scheduling problem (Salmasi *et al.* 2010).
A genetic algorithm for the unrelated parallel machine scheduling problem with sequence dependent setup times (Vallada and Ruiz 2011).
A genetic algorithm for the preemptive and non-preemptive multi-mode resource-constrained project scheduling problem (van Peteghem and Vanhoucke 2010).
A multi-objective genetic algorithm based on immune and entropy principle for flexible job-shop scheduling problem (Wang *et al.* 2010).
A multi-objective ant colony system algorithm for flow shop scheduling problem (Yagmahan and Yenisey 2010).
An efficient ant colony system based on receding horizon control for the aircraft arrival sequencing and scheduling problem (Zhan *et al.* 2010).
An effective genetic algorithm for the flexible job-shop scheduling problem (Zhang *et al.* 2011).
A hybrid immune simulated annealing algorithm for the job shop scheduling problem (Zhang and Wu 2010).

2.7 Shortest Path Problems

A shortest path in a graph defines the shortest possible edges that can be traversed to reach a target node and minimize cost (Mastorakis 2011). Shortest path problems can be classified into Single Source Shortest Path, where the problem of finding a path between two vertices such that the sum of the weights of its constituent edges is minimized (Mastorakis 2011).

The solution to a shortest path problem must contain both information (Daskin 2013): the cost of the shortest path between origin s and destination t and the actual sequence of links (or nodes) that are traversed in going from s to t (Daskin 2013).

Shortest path problem at one short time fraction with an intense terminal traffic conditions and thus, dynamically assigned path nodes for dynamic yard operations (nodes network) can be modeled by a graph $G = (V, E)$ where it comprises a set of vertices or nodes V and a set of E of edges or lines. A tour at the yard area within the dynamically assigned path nodes can be represented via a permutation $\tau = (\tau_1, \tau_2, ..., \tau_n)$. The shortest distance at yard area is formulated as,

Objective function is

$$\text{Minimize } Z = \sum_{(i,j) \in A} C_{ij} x_{ij}$$

Subject to

$$\sum_{\{j:(j,i) \in A\}} x_{ji} - \sum_{\{i:(i,j) \in A\}} x_{ij} = -1 \text{ if } i = s, \; 0 \text{ if } i \neq s \text{ or } d \quad \forall i \in N, \; 1 \text{ if } i = d$$

$$x_{ij} \geq 0 \quad \forall (i,j) \in A$$

where

N   set of number of nodes at seaport terminal (seaside nodes and yard area/stacking area nodes)

A   set of existing arcs (i,j)

C_{ij} arc length (or arc cost) united with each arc (i,j)

i   = s for source node, or i = d for destination node

x_{ij} is the flow from node i to node j

The objective function is to minimize the total distance that is dynamically defined on the seaside and yard area. Constraints ensure that the every point (nodes) visited only once and all these points are included in a tour.

The most recent and highly cited studies about the shortest path problem are shown in Table 2.4.

Table 2.7 Studies in the literature about the shortest path problem

An evolutionary solution for multimodal shortest path problem in metropolises (Abbaspour and Samadzadegan 2010).
An extended shortest path problem: a data envelopment analysis approach (Amirteimoori 2012).
An enhanced exact procedure for the absolute robust shortest path problem (Bruni and Guerriero 2010).
Bicriterion shortest path problem with a general nonadditive cost (Chen and Nie 2013).
Fuzzy dijkstra algorithm for shortest path problem under uncertain environment (Deng et al. 2012).
Tight analysis of the (1+1)-ea for the single source shortest path problem (Doerr et al. 2011).
Solving the fuzzy shortest path problem using multi-criteria decision method based on vague similarity measure (Dou et al. 2012).
New models for the robust shortest path problem: complexity, resolution and generalization (Gabrel et al. 2013).
Shortest path problem with uncertain arc lengths (Gao 2011).
An ant colony optimization algorithm for the bi-objective shortest path problem (Ghoseiri and Nadjari 2010).
Exploring the runtime of an evolutionary algorithm for the multi-objective shortest path problem (Horoba 2010).
The shortest path problem on a fuzzy time-dependent network (Huang and Ding 2012).
An aggregate label setting policy for the multi-objective shortest path problem (Iori et al. 2010).
Solving the constrained shortest path problem using random search strategy (Li et al. 2010).
On an exact method for the constrained shortest path problem (Lozano and Medaglia 2013).
An efficient dynamic model for solving the shortest path problem (Nazemi and Omidi 2013).
On algorithms for the tricriteria shortest path problem with two bottleneck objective functions (Pinto and Pascoal 2010).
A computational study of solution approaches for the resource constrained elementary shortest path problem (Pugliese and Guerriero 2012).
Dynamic programming approaches to solve the shortest path problem with forbidden paths (Pugliese and Guerriero 2013).
Shortest path problem with forbidden paths: the elementary version (Pugliese and Guerriero 2013).
A shortest path problem in an urban transportation network based on driver perceived travel time (Ramazani et al. 2010).
K constrained shortest path problem (Shi 2010).
A simplified algorithm for the all pairs shortest path problem with o(n(2) log n) expected time (Takaoka 2013).
A physarum polycephalum optimization algorithm for the bi-objective shortest path problem (Zhang et al. 2014).
A novel algorithm for all pairs shortest path problem based on matrix multiplication and pulse coupled neural network (Zhang et al. 2011).

2.7.1 Snapshot of an example solved by using Dijkstra's algorithm for the shortest path problem

The solution is depicted in Figure 2.1.

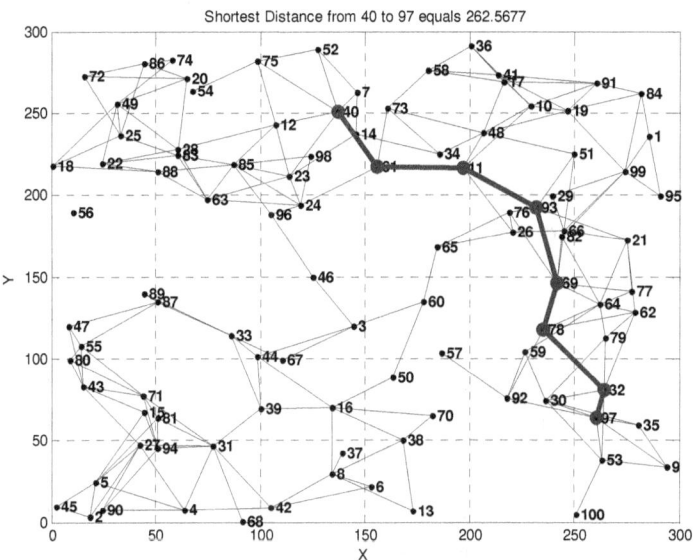

Figure 2.1 The Dijkstra's algortihm's solution for the shortest distance from node 40 to node 97.

Solution parameters are below.

start id = 40

finish id = 97

distance = 262.5677

path = [40 61 11 93 69 78 32 97]

2.8 Travelling Salesman Problem

The traveling salesman problem is a problem in the field of operations research that is reminiscent of Hamiltonian cycles; again, we know of no general solution method (Ore and Wilson, 1990). Suppose a traveling salesman must visit a number of cities before returning home (Ore and Wilson, 1990). Naturally, it is interested in doing this in the shortest possible time, or perhaps he may be concerned to do as cheap as possible (ore and Wilson, 1990).

The traveling salesman problem (TSP) is to find a route of a salesman starting from a location of the house, went to a prescribed set of cities and returns to the original location so that the total distance is minimum and each city is visited exactly once (Gutin and Punnen 2002).

Traveling salesman problem (TSP) is an optimization problem for a given number of cities (e.g., n) and their locations (Yang 2008). Nodes of a graph represent the cities and the distance between the two cities are represented as the weight of an edge or route made between the two cities (Yang 2008). The goal is to find a path that visits each city once and return to the departure city, minimizing the total distance (Yang 2008).

Routing problem I: The problem is to ascertain the operation plan satisfying the demand at various zones at minimum cost.

Objective function is

$$\text{Minimize } f_{obj} = \sum_{i=1}^{G}\sum_{j=1}^{Z}\sum_{k=1}^{F} C_{ijk} x_{ijk}$$

Subject to

$$\sum_{i=1}^{G}\sum_{k=1}^{F} L_k x_{ijk} \geq D_j \qquad \forall j$$

$$\sum_{j=1}^{Z}\sum_{k=1}^{F} L_k x_{ijk} \leq S_j \qquad \forall i$$

$$\sum_{j=1}^{Z} L_k x_{ijk} \leq U_{ki} \qquad \forall k,i$$

$$x_{ijk} \geq 0 \qquad \forall i,j,k$$

where parameters are

G = Number of source locations (index i)

Z = Number of receiving nodes for containers (index j)

F = Number of trailers available (index k)

L_k = Load capacity of trailer k

S_i = Quantity of available containers for transportation from location i

D_j = Quantity of containers required by zone j

C_{ijk} = Unit cost of transporting from location i to zone j by trailer k

U_{ik} = Maximum allowable containers that can be transported from location i by trailer k in a given period

and variables

x_{ijk} = the number of trips required by trailer k from location i to zone j

Routing Problem II: A generic model that practitioners encounter in many planning and decision processes. For instance, the delivery and collection of containers/cargos, etc.

Objective function is

$$\text{Minimize } Z = \sum_{k=1}^{K} \sum_{(i,j) \in A} C_{ij} x_{kij}$$

Subject to

$$\sum_{i=1}^{n} y_{ij} = 1, \quad j = 2, 3, \ldots, n$$

$$\sum_{j=1}^{n} y_{ij} = 1, \quad i = 2, 3, \ldots, n$$

$$\sum_{j=1}^{n} y_{1j} = K$$

$$\sum_{j=1}^{n} y_{i1} = K$$

$$\sum_{i=1}^{n} \sum_{j=2}^{n} D_j x_{kij} \leq U, \quad k = 1, 2, \ldots, K$$

$$\sum_{k=1}^{K} x_{kij} = y_{ij} \quad \forall i, j$$

$$\sum_{(i,j) \in S \times S} y_{ij} \leq |S| - 1, \quad \text{for all subsets } S \text{ of } \{2, 3, \ldots, n\}$$

$$x_{kij} = 0 \text{ or } 1 \quad \forall (i,j) \in A \text{ and } \forall k$$

$$y_{ij} = 0 \text{ or } 1 \quad \forall (i,j) \in A$$

- A fleet of M capacitated vehicles located in a depot (i=1)
- A set of target zones (of size N-1), each having a demand D_j (j=2,...,N)

- A cost C_{ij} of traveling from location i to location j
- The problem is to find a set of routes for delivering / picking up goods to/from the target zones at minimum possible cost.

The vehicle fleet is homogeneous and that each vehicle has a capacity of U units.

and variables:

x_{kij} = 1 if the vehicle k travels on the arc i to j, 0 otherwise

y_{ij} = 1 if any vehicle travels on the arc (i,j), 0 otherwise

The TSP is known to be NP-hard combinatorial optimization problem, which implies that there is no algorithm to solve all cases of problems in polynomial time (Mester *et al.* 2010). Heuristics are often the only alternative for providing high quality, but not necessarily optimal solutions (Mester *et al.* 2010).

The most recent and highly cited studies about the travelling salesman problem are shown in Table 2.8. Various example TSP problems with their optimal solutions are depicted in Figures 2.2 through 2.8.

Table 2.8 Studies in the literature about the traveling salesman problem

Development a new mutation operator to solve the traveling salesman problem by aid of genetic algorithms (Albayrak and Allahverdi 2011).
Amoeba-based neurocomputing for 8-city traveling salesman problem (Aono *et al.* 2011).
Estimation-based metaheuristics for the probabilistic traveling salesman problem (Balaprakash *et al.* 2010).
The traveling salesman problem with pickups, deliveries, and handling costs (Battarra *et al.* 2010).
A memetic algorithm with a large neighborhood crossover operator for the generalized traveling salesman problem (Bontoux *et al.* 2010).
Experimental demonstration of a quantum annealing algorithm for the traveling salesman problem in a nuclear-magnetic-resonance quantum simulator (Chen *et al.* 2011).
Parallelized genetic ant colony systems for solving the traveling salesman problem (Chen *et al.* 2011).
Solving the traveling salesman problem based on the genetic simulated annealing ant colony system with particle swarm optimization techniques (Chen and Chien 2011).
Branch-and-cut for the pickup and delivery traveling salesman problem with fifo loading (Cordeau *et al.* 2010).
A branch-and-cut algorithm for the pickup and delivery traveling salesman problem with lifo loading (Cordeau *et al.* 2010).
The traveling salesman problem with pickup and delivery: polyhedral results and a branch-and-cut algorithm (Dumitrescu *et al.* 2010).
An application of the self-organizing map in the non-euclidean traveling salesman problem (Faigl *et al.* 2011).
Eugenic bacterial memetic algorithm for fuzzy road transport traveling salesman problem (Foldesi *et al.* 2011).
Solving the traveling salesman problem based on an adaptive simulated annealing algorithm with greedy search (Geng *et al.* 2011).
Verification and rectification of the physical analogy of simulated annealing for the solution of the traveling salesman problem (Hasegawa 2011).
A concise guide to the traveling salesman problem (Laporte 2010).
Different initial solution generators in genetic algorithms for solving the probabilistic traveling salesman problem (Liu 2010).
Two-phase pareto local search for the biobjective traveling salesman problem (Lust and Teghem 2010).
A hybrid multi-swarm particle swarm optimization algorithm for the probabilistic traveling salesman problem (Marinakis and Marinaki 2010).
Honey bees mating optimization algorithm for the euclidean traveling salesman problem (Marinakis *et al.* 2011).
Traveling salesman problem heuristics: leading methods, implementations and latest advances (Rego *et al.* 2011).
An ensemble of discrete differential evolution algorithms for solving the generalized traveling salesman problem (Tasgetiren *et al.* 2010).
Solving traveling salesman problem in the adleman-lipton model (Wang *et al.* 2012).
Chaotic ant swarm for the traveling salesman problem (Wei *et al.* 2011).
A parallel immune algorithm for traveling salesman problem and its application on cold rolling scheduling (Zhao *et al.* 2011).

Optimization of Logistics: Theory and Practice

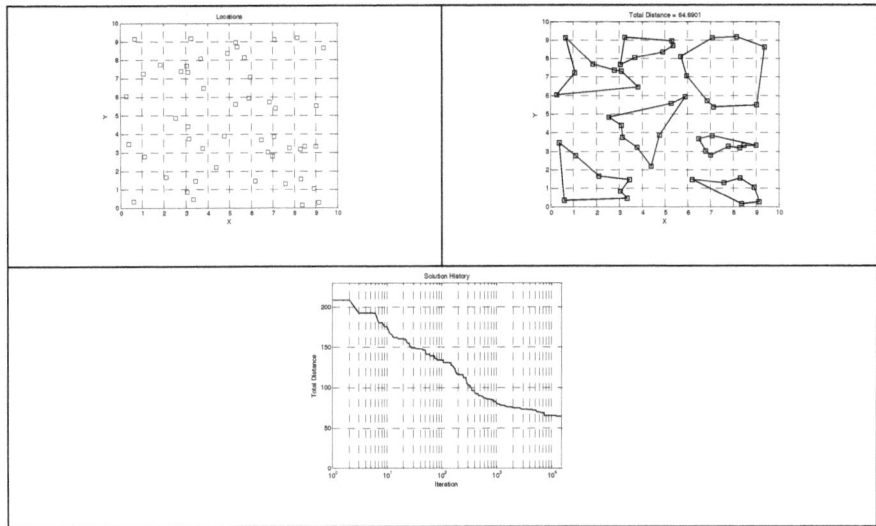

Figure 2.2 Multiple Traveling Salesmen Problem (MTSP) – Each salesman travels to a unique set of cities and completes the route by returning to the city he started from. Each city is visited by exactly one salesman – 7 salesmen, 50 locations

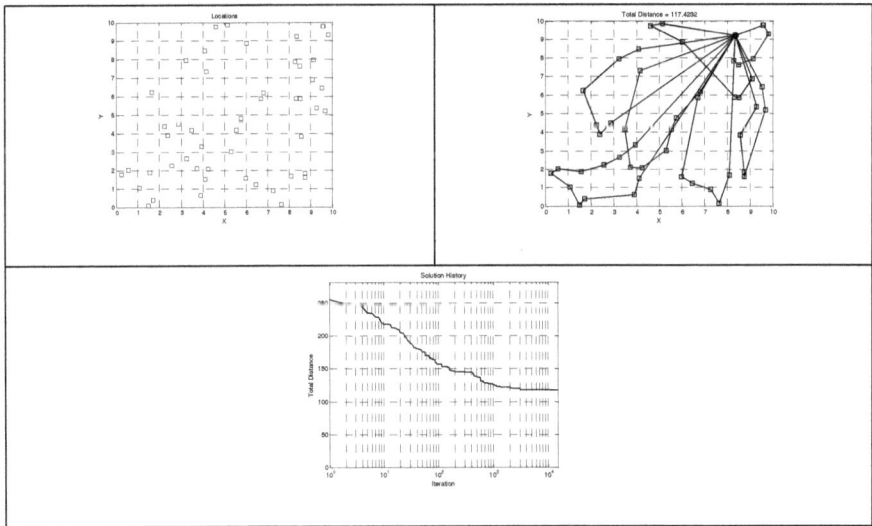

Figure 2.3 Fixed Multiple Traveling Salesmen Problem (MTSPF) – Each salesman starts at the first point, and ends at the first point, but travels to a unique set of cities in between. Except for the first, each city is visited by exactly one salesman – 7 salesmen, 50 locations

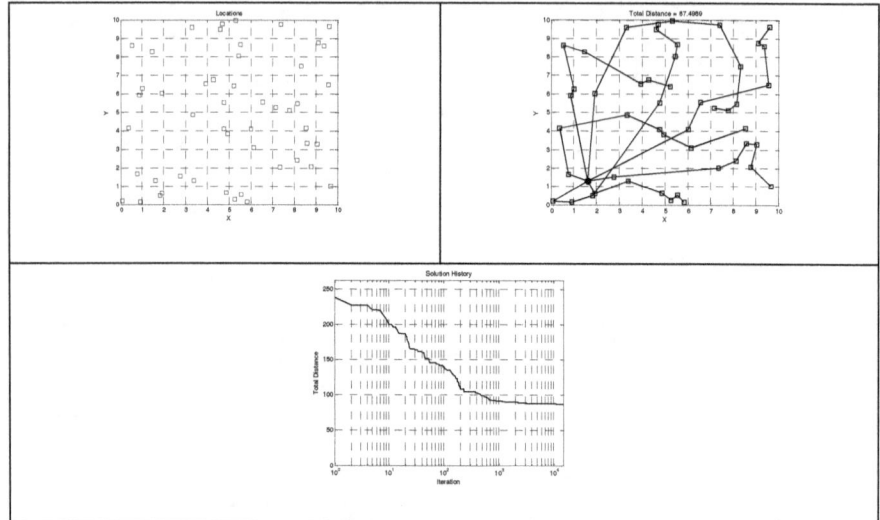

Figure 2.4 Fixed Start Open Multiple Traveling Salesmen Problem (MTSPOFS) - Each salesman starts at the first point and travels to a unique set of cities after that. Except for the first, each city is visited by exactly one salesman - 7 salesmen, 50 locations

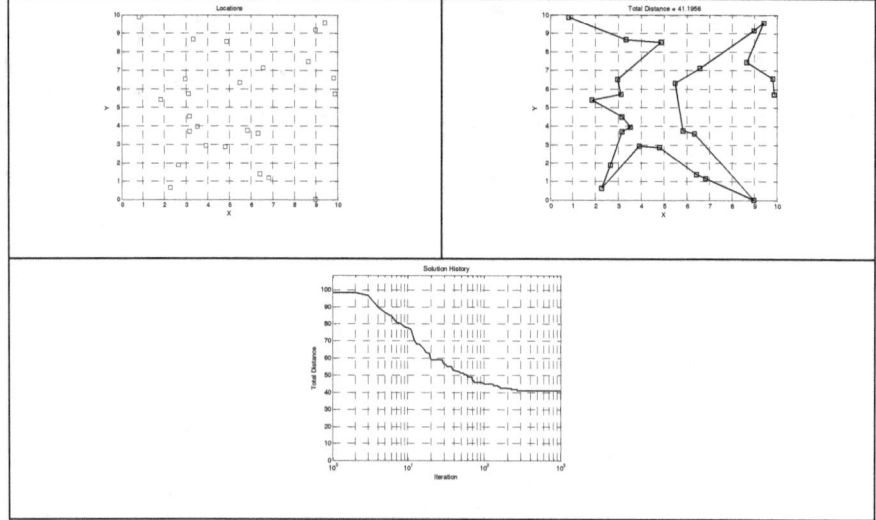

Figure 2.5 Open Traveling Salesman Problem (TSPO) - A single salesman travels to each of the cities but does not close the loop by returning to the city he started from. Each city is visited by the salesman exactly once.

Optimization of Logistics: Theory and Practice

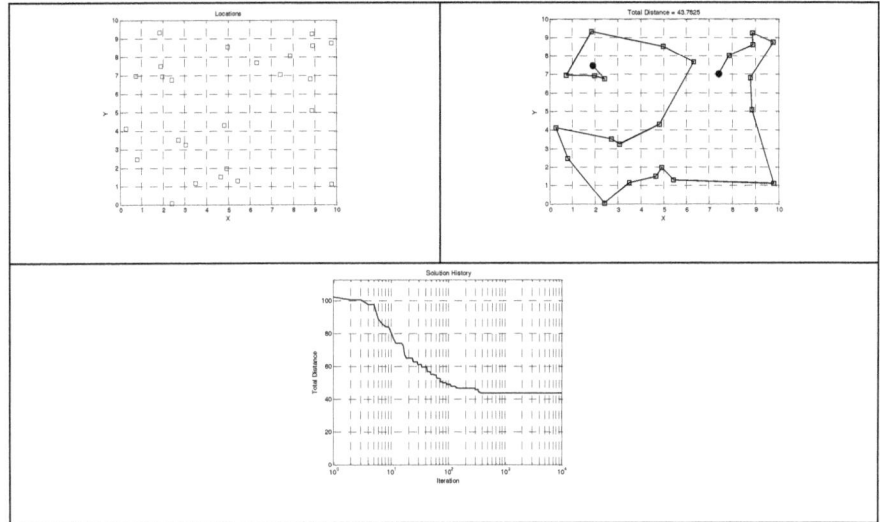

Figure 2.6 Fixed Open Traveling Salesman Problem (TSPOF) - A single salesman starts at the first point, ends at the last point, and travels to each of the remaining cities in between. Each city is visited by the salesman exactly once.

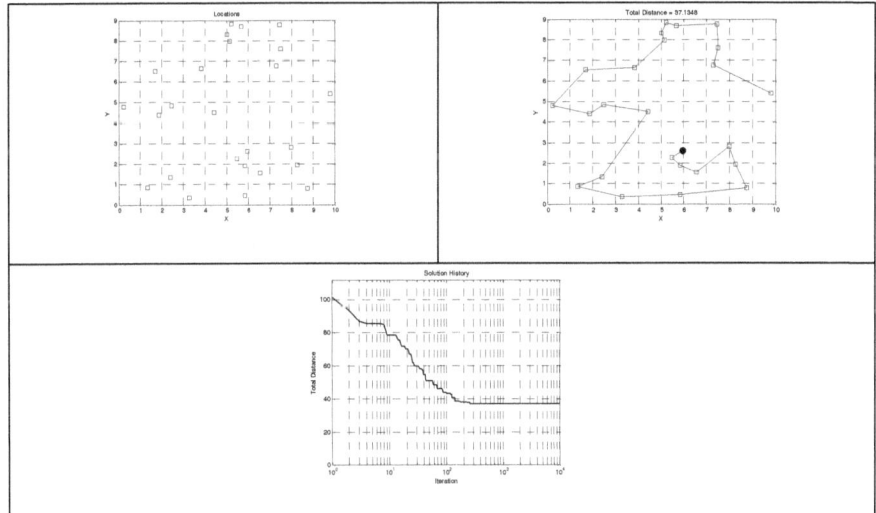

Figure 2.7 Fixed Start Open Traveling Salesman Problem (TSPOFS) - A single salesman starts at the first point and travels to each of the remaining cities but does not close the loop by returning to the city he started from. Each city is visited by the salesman exactly once.

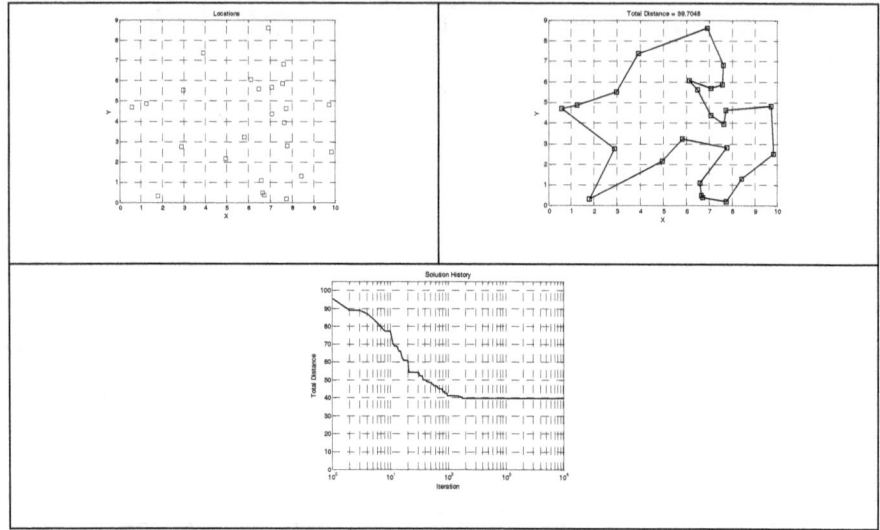

Figure 2.8 Traveling Salesman Problem (TSP) – A single salesman travels to each of the cities and completes the route by returning to the city he started from. Each city is visited by the salesman exactly once.

References

Abbaspour, R. A., & Samadzadegan, F. (2010). An Evolutionary Solution for Multimodal Shortest Path Problem in Metropolises. Computer Science and Information Systems, 7(4), 789-811. doi: 10.2298/csis090710024a

Absi, N., Kedad-Sidhoum, S., & Dauzere-Peres, S. (2011). Uncapacitated lot-sizing problem with production time windows, early productions, backlogs and lost sales. International Journal of Production Research, 49(9), 2551-2566. doi: 10.1080/00207543.2010.532920

Agarwal, A., Colak, S., & Erenguc, S. (2011). A Neurogenetic approach for the resource-constrained project scheduling problem. Computers & Operations Research, 38(1), 44-50. doi: 10.1016/j.cor.2010.01.007

Albareda-Sambola, M., Alonso-Ayuso, A., Escudero, L. F., Fernandez, E., Hinojosa, Y., & Pizarro-Romero, C. (2010). A computational comparison of several formulations for the multi-period incremental service facility location problem. Top, 18(1), 62-80. doi: 10.1007/s11750-009-0106-3

Albareda-Sambola, M., Fernandez, E., & Saldanha-da-Gama, F. (2011). The facility location problem with Bernoulli demands. Omega-International Journal of Management Science, 39(3), 335-345. doi: 10.1016/j.omega.2010.08.002

Albayrak, M., & Allahverdi, N. (2011). Development a new mutation operator to solve the Traveling Salesman Problem by aid of Genetic Algorithms. Expert Systems with Applications, 38(3), 1313-1320. doi: 10.1016/j.eswa.2010.07.006

Amirteimoori, A. (2012). An extended shortest path problem: A data envelopment analysis approach. Applied Mathematics Letters, 25(11), 1839-1843. doi: 10.1016/j.aml.2012.02.042

Angelelli, E., Mansini, R., & Speranza, M. G. (2010). Kernel search: A general heuristic for the multi-dimensional knapsack problem. Computers & Operations Research, 37(11) 2017-2026. doi: 10.1016/j.cor.2010.02.002

Aono, M., Zhu, L. P., & Hara, M. (2011). Amoeba-based Neurocomputing for 8-City Traveling Salesman Problem. International Journal of Unconventional Computing, 7(6), 463-480.

Argyris, N., Figueira, J. R., & Morton, A. (2011). Identifying preferred solutions to Multi-Objective Binary Optimisation problems, with an application to the Multi-Objective Knapsack Problem. Journal of Global Optimization, 49(2), 213-235. doi: 10.1007/s10898-010-9541-9

Aros-Vera, F., Marianov, V., & Mitchell, J. E. (2013). p-Hub approach for the optimal park-and-ride facility location problem. European Journal of Operational Research, 226(2), 277-285. doi: 10.1016/j.ejor.2012.11.006

Avella, P., Boccia, M., & Vasilyev, I. (2010). A computational study of exact knapsack separation for the generalized assignment problem. Computational Optimization and Applications, 45(3), 543-555. doi: 10.1007/s10589-008-9183-8

Azi, N., Gendreau, M., & Potvin, J. Y. (2010). An exact algorithm for a vehicle routing problem with time windows and multiple use of vehicles. European Journal of Operational Research 202(3), 756-763. doi: 10.1016/j.ejor.2009.06.034

Bagheri, A., Zandieh, M., Mahdavi, I., & Yazdani, M. (2010). An artificial immune algorithm for the flexible job-shop scheduling problem. Future Generation Computer Systems-the International Journal of Grid Computing and Escience, 26(4), 533-541. doi: 10.1016/j.future.2009.10.004

Balaprakash, P., Birattari, M., Stutzle, T., & Dorigo, M. (2010). Estimation-based metaheuristics for the probabilistic traveling salesman problem. Computers & Operations Research, 37(11), 1939-1951. doi: 10.1016/j.cor.2009.12.005

Baldacci, R., Bartolini, E., & Laporte, G. (2010). Some applications of the generalized vehicle routing problem. Journal of the Operational Research Society, 61(7), 1072-1077. doi: 10.1057/jors.2009.51

Baldacci, R., Mingozzi, A., & Roberti, R. (2011). New Route Relaxation and Pricing Strategies for the Vehicle Routing Problem. Operations Research, 59(5), 1269-1283. doi: 10.1287/opre.1110.0975

Battarra, M., Erdogan, G., Laporte, G., & Vigo, D. (2010). The Traveling Salesman Problem with Pickups, Deliveries, and Handling Costs. Transportation Science, 44(3), 383-399. doi: 10.1287/trsc.1100.0316

Beltran-Royo, C., Vial, J. P., & Alonso-Ayuso, A. (2012). Semi-Lagrangian relaxation applied to the uncapacitated facility location problem. Computational Optimization and Applications, 51(1), 387-409. doi: 10.1007/s10589-010-9338-2

Benjamin, A. M., & Beasley, J. E. (2010). Metaheuristics for the waste collection vehicle routing problem with time windows, driver rest period and multiple disposal facilities. Computers & Operations Research, 37(12), 2270-2280. doi: 10.1016/j.cor.2010.03.019

Bhattacharya, R., & Bandyopadhyay, S. (2010). Solving conflicting bi-objective facility location problem by NSGA II evolutionary algorithm. International Journal of Advanced Manufacturing Technology, 51(1-4), 397-414. doi: 10.1007/s00170-010-2622-6

Bontoux, B., Artigues, C., & Feillet, D. (2010). A Memetic Algorithm with a large neighborhood crossover operator for the Generalized Traveling Salesman Problem. Computers & Operations Research, 37(11), 1844-1852. doi: 10.1016/j.cor.2009.05.004

Boussier, S., Vasquez, M., Vimont, Y., Hanafi, S., & Michelon, P. (2010). A multi-level search strategy for the 0-1 Multidimensional Knapsack Problem. Discrete Applied Mathematics, 158(2), 97-109. doi: 10.1016/j.dam.2009.08.007

Bowman, M., Briand, L. C., & Labiche, Y. (2010). Solving the Class Responsibility Assignment Problem in Object-Oriented Analysis with Multi-Objective Genetic Algorithms. Ieee Transactions on Software Engineering, 36(6), 817-837. doi: 10.1109/tse.2010.70

Bramel, Julien; Simchi-Levi, David. Logic of Logistics : Theory, Algorithms, and Applications for Logistics Management. Secaucus, NJ, USA: Springer, 1997. p 203.

Brandao, J. (2011). A tabu search algorithm for the heterogeneous fixed fleet vehicle routing problem. Computers & Operations Research, 38(1), 140-151. doi: 10.1016/j.cor.2010.04.008

Bruni, M. E., & Guerriero, F. (2010). An enhanced exact procedure for the absolute robust shortest path problem. International Transactions in Operational Research, 17(2) 207-220. doi: 10.1111/j.1475-3995.2009.00702.x

Budish, E., & Cantillon, E. (2012). The Multi-unit Assignment Problem: Theory and Evidence from Course Allocation at Harvard. American Economic Review, 102(5), 2237-2271. doi: 10.1257/aer.102.5.2237

Byrka, J., & Aardal, K. (2010). An Optimal Bifactor Approximation Algorithm For The Metric Uncapacitated Facility Location Problem. Siam Journal on Computing, 39(6), 2212-2231. doi: 10.1137/070708901

Camargo, V. C. B., Mattiolli, L., & Toledo, F. M. B. (2012). A knapsack problem as a tool to solve the production planning problem in small foundries. Computers & Operations Research, 39(1), 86-92. doi: 10.1016/j.cor.2010.10.023

Cardenas-Barron, L. E. (2010). Adaptive genetic algorithm for lot-sizing problem with self-adjustment operation rate: A discussion. International Journal of Production Economics, 123(1), 243-245. doi: 10.1016/j.ijpe.2009.07.007

Carrizosa, E., Ushakov, A., & Vasilyev, I. (2012). A computational study of a nonlinear minsum facility location problem. Computers & Operations Research, 39(11), 2625-2633. doi: 10.1016/j.cor.2012.01.009

Cela, E., Schmuck, N. S., Wimer, S., & Woeginger, G. J. (2011). The Wiener maximum quadratic assignment problem. Discrete Optimization, 8(3), 411-416. doi: 10.1016/j.disopt.2011.02.002

Chen, B. Y., Lam, W. H. K., Sumalee, A., & Shao, H. (2011). An efficient solution algorithm for solving multi-class reliability-based traffic assignment problem. Mathematical and Computer Modelling, 54(5-6), 1428-1439. doi: 10.1016/j.mcm.2011.04.015

Chen, Der-San; Batson, Robert G.; Dang, Yu. Applied Integer Programming Modeling and Solution. Hoboken, NJ, USA: Wiley 2011. p 30-345.

Chen, H. W., Kong, X., Chong, B., Qin, G., Zhou, X. Y., Peng, X. H., & Du, J. F. (2011). Experimental demonstration of a quantum annealing algorithm for the traveling salesman problem in a nuclear-magnetic-resonance quantum simulator. Physical Review A, 83(3). doi: 10.1103/PhysRevA.83.032314

Chen, L., & Lu, Z. Q. (2012). The storage location assignment problem for outbound containers in a maritime terminal. International Journal of Production Economics, 135(1), 73-80. doi: 10.1016/j.ijpe.2010.09.019

Chen, P., & Nie, Y. (2013). Bicriterion shortest path problem with a general nonadditive cost. Transportation Research Part B-Methodological, 57, 419-435. doi: 10.1016/j.trb.2013.05.008

Chen, P., Huang, H. K., & Dong, X. Y. (2010). Iterated variable neighborhood descent algorithm for the capacitated vehicle routing problem. Expert Systems with Applications, 37(2), 1620-1627. doi: 10.1016/j.eswa.2009.06.047

Chen, S. M., & Chien, C. Y. (2011). Parallelized genetic ant colony systems for solving the traveling salesman problem. Expert Systems with Applications, 38(4), 3873-3883. doi: 10.1016/j.eswa.2010.09.048

Chen, S. M., & Chien, C. Y. (2011). Solving the traveling salesman problem based on the genetic simulated annealing ant colony system with particle swarm optimization techniques. Expert

Systems with Applications, 38(12), 14439-14450. doi: 10.1016/j.eswa.2011.04.163

Cheng, T. C. E., Cheng, S. R., Wu, W. H., Hsu, P. H., & Wu, C. C. (2011). A two-agent single-machine scheduling problem with truncated sum-of-processing-times-based learning considerations. Computers & Industrial Engineering, 60(4), 534-541. doi: 10.1016/j.cie.2010.12.008

Cherfi, N., & Hifi, M. (2010). A column generation method for the multiple-choice multi-dimensional knapsack problem. Computational Optimization and Applications, 46(1), 51-73. doi: 10.1007/s10589-008-9184-7

Chvátal, V. (Editor). Combinatorial Optimization : Methods and Applications. Amsterdam, NLD: IOS Press 2011. p 97.

Cordeau, J. F., Dell'Amico, M., & Iori, M. (2010). Branch-and-cut for the pickup and delivery traveling salesman problem with FIFO loading. Computers & Operations Research, 37(5), 970-980. doi: 10.1016/j.cor.2009.08.003

Cordeau, J. F., Iori, M., Laporte, G., & Gonzalez, J. J. S. (2010). A Branch-and-Cut Algorithm for the Pickup and Delivery Traveling Salesman Problem with LIFO Loading. Networks, 55(1), 46-59. doi: 10.1002/net.20312

D'Acierno, L., Gallo, M., & Montella, B. (2012). An Ant Colony Optimisation algorithm for solving the asymmetric traffic assignment problem. European Journal of Operational Research, 217(2), 459-469. doi: 10.1016/j.ejor.2011.09.035

Das, D., Roy, A., & Kar, S. (2011). A volume flexible economic production lot-sizing problem with imperfect quality and random machine failure in fuzzy-stochastic environment. Computers & Mathematics with Applications, 61(9), 2388-2400. doi: 10.1016/j.camwa.2011.02.015

Daskin, Mark S.. Network and Discrete Location : Models, Algorithms, and Applications (2nd Edition). Somerset, NJ, USA: Wiley 2013. p 3-66.

Davis, L., Samanlioglu, F., Jiang, X. C., Mota, D., & Stanfield, P. (2012). A heuristic approach for allocation of data to RFID tags: A data allocation knapsack problem (DAKP). Computers & Operations Research, 39(1), 93-104. doi: 10.1016/j.cor.2011.01.019

De Giovanni, L., & Pezzella, F. (2010). An Improved Genetic Algorithm for the Distributed and Flexible Job-shop Scheduling problem. European Journal of Operational Research 200(2), 395-

408. doi: 10.1016/j.ejor.2009.01.008

de Klerk, E., & Sotirov, R. (2010). Exploiting group symmetry in semidefinite programming relaxations of the quadratic assignment problem. Mathematical Programming, 122(2), 225-246. doi: 10.1007/s10107-008-0246-5

Deng, Y., Chen, Y. X., Zhang, Y. J., & Mahadevan, S. (2012). Fuzzy Dijkstra algorithm for shortest path problem under uncertain environment. Applied Soft Computing, 12(3), 1231-1237. doi: 10.1016/j.asoc.2011.11.011

Desaulniers, G. (2010). Branch-and-Price-and-Cut for the Split-Delivery Vehicle Routing Problem with Time Windows. Operations Research, 58(1), 179-192. doi: 10.1287/opre.1090.0713

Diabat, A., Abdallah, T., Al-Refaie, A., Svetinovic, D., & Govindan, K. (2013). Strategic Closed-Loop Facility Location Problem With Carbon Market Trading. Ieee Transactions on Engineering Management, 60(2), 398-408. doi: 10.1109/tem.2012.2211105

Dizdar, D., Gershkov, A., & Moldovanu, B. (2011). Revenue maximization in the dynamic knapsack problem. Theoretical Economics, 6(2), 157-184. doi: 10.3982/te700

Doerr, B., Happ, E., & Klein, C. (2011). Tight Analysis of the (1+1)-EA for the Single Source Shortest Path Problem. Evolutionary Computation, 19(4), 673-691.

Dou, Y. L., Zhu, L. C., & Wang, H. S. (2012). Solving the fuzzy shortest path problem using multi-criteria decision method based on vague similarity measure. Applied Soft Computing, 12(6), 1621-1631. doi: 10.1016/j.asoc.2012.03.013

Du, D. L., Lu, R. X., & Xu, D. C. (2012). A Primal-Dual Approximation Algorithm for the Facility Location Problem with Submodular Penalties. Algorithmica, 63(1-2), 191-200. doi: 10.1007/s00453-011-9526-1

Du, D. L., Wang, X., & Xu, D. C. (2010). An approximation algorithm for the k-level capacitated facility location problem. Journal of Combinatorial Optimization 20(4), 361-368. doi: 10.1007/s10878-009-9213-1

Dugardin, F., Yalaoui, F., & Amodeo, L. (2010). New multi-objective method to solve reentrant hybrid flow shop scheduling problem. European Journal of Operational Research 203(1), 22-31. doi: 10.1016/j.ejor.2009.06.031

Duhamel, C., Lacomme, P., Quilliot, A., & Toussaint, H. (2011). A multi-start evolutionary local search for the two-dimensional loading capacitated vehicle routing problem. Computers & Operations Research, 38(3), 617-640. doi: 10.1016/j.cor.2010.08.017

Dumitrescu, I., Ropke, S., Cordeau, J. F., & Laporte, G. (2010). The traveling salesman problem with pickup and delivery: polyhedral results and a branch-and-cut algorithm. Mathematical Programming, 121(2), 269-305. doi: 10.1007/s10107-008-0234-9

Dye, C. Y., & Hsieh, T. P. (2010). A particle swarm optimization for solving joint pricing and lot-sizing problem with fluctuating demand and unit purchasing cost. Computers & Mathematics with Applications, 60(7), 1895-1907. doi: 10.1016/j.camwa.2010.07.023

Dye, C. Y., & Ouyang, L. Y. (2011). A particle swarm optimization for solving joint pricing and lot-sizing problem with fluctuating demand and trade credit financing. Computers & Industrial Engineering, 60(1), 127-137. doi: 10.1016/j.cie.2010.10.010

El-Rewini, Hesham; Abd-El-Barr, Mostafa. Advanced Computer Architecture and Parallel Processing. Hoboken, NJ, USA: Wiley 2005. p 251-252.

Faigl, J., Kulich, M., Vonasek, V., & Preucil, L. (2011). An application of the self-organizing map in the non-Euclidean Traveling Salesman Problem. Neurocomputing, 74(5), 671-679. doi: 10.1016/j.neucom.2010.08.026

Feng, Y., Chen, S. X., Kumar, A., & Lin, B. (2011). Solving single-product economic lot-sizing problem with non-increasing setup cost, constant capacity and convex inventory cost in $O(N \log N)$ time. Computers & Operations Research, 38(4), 717-722. doi: 10.1016/j.cor.2010.08.009

Figliozzi, M. A. (2010). An iterative route construction and improvement algorithm for the vehicle routing problem with soft time windows. Transportation Research Part C-Emerging Technologies, 18(5), 668-679. doi: 10.1016/j.trc.2009.08.005

Foldesi, P., Botzheim, J., & Koczy, L. T. (2011). Eugenic Bacterial Memetic Algorithm For Fuzzy Road Transport Traveling Salesman Problem. International Journal of Innovative Computing Information and Control, 7(5B), 2775-2798.

Gabor, A. F., & van Ommeren, J. (2010). A new approximation algorithm for the multilevel facility location problem. Discrete Applied Mathematics, 158(5), 453-460. doi: 10.1016/j.dam.2009.11.007

Gabrel, V., Murat, C., & Wu, L. (2013). New models for the robust shortest path problem: complexity, resolution and generalization.

Annals of Operations Research 207(1), 97-120. doi: 10.1007/s10479-011-1004-2

Gao, Y. (2011). Shortest path problem with uncertain arc lengths. Computers & Mathematics with Applications, 62(6), 2591-2600. doi: 10.1016/j.camwa.2011.07.058

Garcia-Najera, A., & Bullinaria, J. A. (2011). An improved multi-objective evolutionary algorithm for the vehicle routing problem with time windows. Computers & Operations Research, 38(1), 287-300. doi: 10.1016/j.cor.2010.05.004

Geng, X. T., Chen, Z. H., Yang, W., Shi, D. Q., & Zhao, K. (2011). Solving the traveling salesman problem based on an adaptive simulated annealing algorithm with greedy search. Applied Soft Computing, 11(4), 3680-3689. doi: 10.1016/j.asoc.2011.01.039

Ghasemi, T., & Razzazi, M. (2011). Development of core to solve the multidimensional multiple-choice knapsack problem. Computers & Industrial Engineering, 60(2), 349-360. doi: 10.1016/j.cie.2010.12.001

Ghiani, Gianpaolo. Wiley Essentials in Operations Research and Management Science : Introduction to Logistics Systems Management (2nd Edition). Somerset, NJ, USA: Wiley 2013. p 146-342.

Ghoseiri, K., & Nadjari, B. (2010). An ant colony optimization algorithm for the bi-objective shortest path problem. Applied Soft Computing, 10(4), 1237-1246. doi: 10.1016/j.asoc.2009.09.014

Godrich, H., Petropulu, A. P., & Poor, H. V. (2012). Sensor Selection in Distributed Multiple-Radar Architectures for Localization: A Knapsack Problem Formulation. Ieee Transactions on Signal Processing, 60(1), 247-260. doi: 10.1109/tsp.2011.2170170

Gonzalez-Ramirez, R. G., Smith, N. R., & Askin, R. G. (2011). A heuristic approach for a multi-product capacitated lot-sizing problem with pricing. International Journal of Production Research, 49(4), 1173-1196. doi: 10.1080/00207540903524482

Goyal, V., & Ravi, R. (2010). A PTAS for the chance-constrained knapsack problem with random item sizes. Operations Research Letters, 38(3), 161-164. doi: 10.1016/j.orl.2010.01.003

Gu, J. W., Gu, M. Z., Cao, C. W., & Gu, X. S. (2010). A novel competitive co-evolutionary quantum genetic algorithm for stochastic job shop scheduling problem. Computers & Operations Research, 37(5), 927-937. doi: 10.1016/j.cor.2009.07.002

Guan, Y. P., & Liu, T. M. (2010). Stochastic lot-sizing problem with inventory-bounds and constant order-capacities. European Journal of Operational Research 207(3), 1398-1409. doi: 10.1016/j.ejor.2010.07.003

Guillaume, R., Kobylanski, P., & Zielinski, P. (2012). A robust lot sizing problem with ill-known demands. Fuzzy Sets and Systems 206, 39-57. doi: 10.1016/j.fss.2012.01.015

Gutin, Gregory; Punnen, Abraham P.. Travelling Salesman Problem and Its Variations. Secaucus, NJ, USA: Kluwer Academic Publishers 2002. p 20.

Hachicha, W. (2011). A Simulation Metamodelling Based Neural Networks For Lot-Sizing Problem In Mto Sector. International Journal of Simulation Modelling, 10(4), 191-203. doi: 10.2507/ijsimm10(4)3.188

Han, X. L., Lu, Z. Q., & Xi, L. F. (2010). A proactive approach for simultaneous berth and quay crane scheduling problem with stochastic arrival and handling time. European Journal of Operational Research 207(3), 1327-1340. doi: 10.1016/j.ejor.2010.07.018

Hanafi, S., & Wilbaut, C. (2011). Improved convergent heuristics for the 0-1 multidimensional knapsack problem. Annals of Operations Research, 183(1), 125-142. doi: 10.1007/s10479-009-0546-z

Hartmann, S., & Briskorn, D. (2010). A survey of variants and extensions of the resource-constrained project scheduling problem. European Journal of Operational Research 207(1), 1-14. doi: 10.1016/j.ejor.2009.11.005

Hasegawa, M. (2011). Verification and rectification of the physical analogy of simulated annealing for the solution of the traveling salesman problem. Physical Review E, 83(3). doi: 10.1103/PhysRevE.83.036708

Helber, S., & Sahling, F. (2010). A fix-and-optimize approach for the multi-level capacitated lot sizing problem. International Journal of Production Economics, 123(2), 247-256. doi: 10.1016/j.ijpe.2009.08.022

Hill, R. R., Cho, Y. K., & Moore, J. T. (2012). Problem reduction heuristic for the 0-1 multidimensional knapsack problem. Computers & Operations Research, 39(1), 19-26. doi: 10.1016/j.cor.2010.06.009

Horoba, C. (2010). Exploring the Runtime of an Evolutionary Algorithm for the Multi-Objective Shortest Path Problem. Evolutionary Computation, 18(3), 357-381. doi: 10.1162/EVCO_a_00014

Huang, W., & Ding, L. X. (2012). The Shortest Path Problem on a Fuzzy Time-Dependent Network. Ieee Transactions on Communications, 60(11), 3376-3385. doi: 10.1109/tcomm.2012.090512.100570

Iori, M., Martello, S., & Pretolani, D. (2010). An aggregate label setting policy for the multi-objective shortest path problem. European Journal of Operational Research 207(3), 1489-1496. doi: 10.1016/j.ejor.2010.06.035

Iyigun, C., & Ben-Israel, A. (2010). A generalized Weiszfeld method for the multi-facility location problem. Operations Research Letters, 38(3) 207-214. doi: 10.1016/j.orl.2009.11.005

Jarboui, Bassem (Editor); Siarry, Patrick (Editor); Teghem, Jacques (Editor). Metaheuristics for Production Scheduling. Somerset, NJ, USA: Wiley-ISTE 2013. p 373-433.

Kalcsics, J., Nickel, S., Puerto, J., & Rodriguez-Chia, A. (2010). The ordered capacitated facility location problem. Top, 18(1) 203-222. doi: 10.1007/s11750-009-0089-0

Ke, L. J., Feng, Z. R., Ren, Z. G., & Wei, X. L. (2010). An ant colony optimization approach for the multidimensional knapsack problem. Journal of Heuristics, 16(1), 65-83. doi: 10.1007/s10732-008-9087-x

Kellerer, H., & Strusevich, V. A. (2010). Fully Polynomial Approximation Schemes for a Symmetric Quadratic Knapsack Problem and its Scheduling Applications. Algorithmica, 57(4), 769-795. doi: 10.1007/s00453-008-9248-1

Khaloozadeh, H., & Baromand, S. (2010). State covariance assignment problem. Iet Control Theory and Applications, 4(3), 391-402. doi: 10.1049/iet-cta.2008.0359

Konstantinidis, A., Yang, K., Zhang, Q. F., & Zeinalipour-Yazti, D. (2010). A multi-objective evolutionary algorithm for the deployment and power assignment problem in wireless sensor networks. Computer Networks, 54(6), 960-976. doi: 10.1016/j.comnet.2009.08.010

Kosuch, S., & Lisser, A. (2010). Upper bounds for the 0-1 stochastic knapsack problem and a B&B algorithm. Annals of Operations Research, 176(1), 77-93. doi: 10.1007/s10479-009-0577-5

Kucukaydin, H., Aras, N., & Altinel, I. K. (2011). Competitive facility location problem with attractiveness adjustment of the follower A bilevel programming model and its solution. European Journal of Operational Research 208(3) 206-220. doi: 10.1016/j.ejor.2010.08.009

Kucukdeniz, T., Baray, A., Ecerkale, K., & Esnaf, S. (2012). Integrated use of fuzzy c-means and convex programming for capacitated multi-facility location problem. Expert Systems with Applications, 39(4), 4306-4314. doi: 10.1016/j.eswa.2011.09.102

Kumar, R., & Singh, P. K. (2010). Assessing solution quality of biobjective 0-1 knapsack problem using evolutionary and heuristic algorithms. Applied Soft Computing, 10(3), 711-718. doi: 10.1016/j.asoc.2009.08.037

Kuo, Y. Y. (2010). Using simulated annealing to minimize fuel consumption for the time-dependent vehicle routing problem. Computers & Industrial Engineering, 59(1), 157-165. doi: 10.1016/j.cie.2010.03.012

Laporte, G. (2010). A concise guide to the Traveling Salesman Problem. Journal of the Operational Research Society, 61(1), 35-40. doi: 10.1057/jors.2009.76

Lau, H. C. W., Chan, T. M., Tsui, W. T., & Pang, W. K. (2010). Application of Genetic Algorithms to Solve the Multidepot Vehicle Routing Problem. Ieee Transactions on Automation Science and Engineering, 7(2), 383-392. doi: 10.1109/tase.2009.2019265

Lee, C. Y., Lee, Z. J., Lin, S. W., & Ying, K. C. (2010). An enhanced ant colony optimization (EACO) applied to capacitated vehicle routing problem. Applied Intelligence, 32(1), 88-95. doi: 10.1007/s10489-008-0136-9

Lee, W. C., Shiau, Y. R., Chen, S. K., & Wu, C. C. (2010). A two-machine flowshop scheduling problem with deteriorating jobs and blocking. International Journal of Production Economics, 124(1), 188-197. doi: 10.1016/j.ijpe.2009.11.001

Lee, W. C., Wang, W. J., Shiau, Y. R., & Wu, C. C. (2010). A single-machine scheduling problem with two-agent and deteriorating jobs. Applied Mathematical Modelling, 34(10), 3098-3107. doi: 10.1016/j.apm.2010.01.015

Lee, W. C., Wu, C. C., & Hsu, P. H. (2010). A single-machine learning effect scheduling problem with release times. Omega-International Journal of Management Science, 38(1-2), 3-11. doi: 10.1016/j.omega.2009.01.001

Letocart, L., Nagih, A., & Plateau, G. (2012). Reoptimization in Lagrangian methods for the 0-1 quadratic knapsack problem. Computers & Operations Research, 39(1), 12-18. doi: 10.1016/j.cor.2010.10.027

Li, J. Q., Pan, Q. K., Suganthan, P. N., & Chua, T. J. (2011). A hybrid tabu search algorithm with an efficient neighborhood structure for the flexible job shop scheduling problem. International Journal of Advanced Manufacturing Technology, 52(5-8), 683-697. doi: 10.1007/s00170-010-2743-y

Li, K. P., Gao, Z. Y., Tang, T., & Yang, L. X. (2010). Solving the constrained shortest path problem using random search strategy. Science China-Technological Sciences, 53(12), 3258-3263. doi: 10.1007/s11431-010-4105-2

Li, X. Y., Baki, F., Tian, P., & Chaouch, B. A. (2014). A robust block-chain based tabu search algorithm for the dynamic lot sizing problem with product returns and remanufacturing. Omega-International Journal of Management Science, 42(1), 75-87. doi: 10.1016/j.omega.2013.03.003

Lin, Y. K., & Yeh, C. T. (2011). Reliability Optimization Of Component Assignment Problem For A Multistate Network In Terms Of Minimal Cuts. Journal of Industrial and Management Optimization, 7(1), 211-227. doi: 10.3934/jimo.2011.7.211

Liu, Y. H. (2010). Different initial solution generators in genetic algorithms for solving the probabilistic traveling salesman problem. Applied Mathematics and Computation, 216(1), 125-137. doi: 10.1016/j.amc.2010.01.021

Lozano, L., & Medaglia, A. L. (2013). On an exact method for the constrained shortest path problem. Computers & Operations Research, 40(1), 378-384. doi: 10.1016/j.cor.2012.07.008

Lu, S., & Nie, Y. (2010). Stability of user-equilibrium route flow solutions for the traffic assignment problem. Transportation Research Part B-Methodological, 44(4), 609-617. doi: 10.1016/j.trb.2009.09.003

Lu, Z. Q., Zhang, Y. J., & Han, X. L. (2013). Integrating run-based preventive maintenance into the capacitated lot sizing problem with reliability constraint. International Journal of Production Research, 51(5), 1379-1391. doi: 10.1080/00207543.2012.693637

Lust, T., & Teghem, J. (2010). Two-phase Pareto local search for the biobjective traveling salesman problem. Journal of Heuristics, 16(3), 475-510. doi: 10.1007/s10732-009-9103-9

Maric, M. (2010). An Efficient Genetic Algorithm For Solving The Multi-Level Uncapacitated Facility Location Problem. Computing and Informatics, 29(2), 183-201.

Marin, A. (2011). The discrete facility location problem with balanced allocation of customers. European Journal of Operational Research,

210(1), 27-38. doi: 10.1016/j.ejor.2010.10.012

Marinakis, Y., & Marinaki, M. (2010). A hybrid genetic - Particle Swarm Optimization Algorithm for the vehicle routing problem. Expert Systems with Applications, 37(2), 1446-1455. doi: 10.1016/j.eswa.2009.06.085

Marinakis, Y., & Marinaki, M. (2010). A Hybrid Multi-Swarm Particle Swarm Optimization algorithm for the Probabilistic Traveling Salesman Problem. Computers & Operations Research, 37(3), 432-442. doi: 10.1016/j.cor.2009.03.004

Marinakis, Y., Marinaki, M., & Dounias, G. (2010). A hybrid particle swarm optimization algorithm for the vehicle routing problem. Engineering Applications of Artificial Intelligence, 23(4), 463-472. doi: 10.1016/j.engappai.2010.02.002

Marinakis, Y., Marinaki, M., & Dounias, G. (2011). Honey bees mating optimization algorithm for the Euclidean traveling salesman problem. Information Sciences, 181(20), 4684-4698. doi: 10.1016/j.ins.2010.06.032

Mastorakis, Nikos E.. Computer Science, Technology and Applications : Pathway Modeling and Algorithm Research. New York, NY, USA: Nova Science Publishers, Inc. 2011. p 46.

Mateus, G. R., Resende, M. G. C., & Silva, R. M. A. (2011). GRASP with path-relinking for the generalized quadratic assignment problem. Journal of Heuristics, 17(5), 527-565. doi: 10.1007/s10732-010-9144-0

McLay, L. A., Lloyd, J. D., & Niman, E. (2011). Interdicting nuclear material on cargo containers using knapsack problem models. Annals of Operations Research, 187(1), 185-205. doi: 10.1007/s10479-009-0667-4

Mendoza, J. E., Castanier, B., Gueret, C., Medaglia, A. L., & Velasco, N. (2010). A memetic algorithm for the multi-compartment vehicle routing problem with stochastic demands. Computers & Operations Research, 37(11), 1886-1898. doi: 10.1016/j.cor.2009.06.015

Menlo Park, CA, USA: Course Technology / Cengage Learning, 1996. p 51-56.

Mester, D. (Editor); Ronin, D. (Editor); Frenkel, M. (Editor). Genetics - Research and Issues : Discrete Optimization for Some TSP-like Genome Mapping Problems. New York, NY, USA: Nova Science Publishers, Inc. 2010. p 6.

Mishra, D.N.; Agarwal, S.K.. Operation Research. Lucknow, IND:

Global Media 2009. p 107.

Mohammadi, M., Torabi, S. A., Ghomi, S., & Karimi, B. (2010). A new algorithmic approach for capacitated lot-sizing problem in flow shops with sequence-dependent setups. International Journal of Advanced Manufacturing Technology, 49(1-4) 201-211. doi: 10.1007/s00170-009-2366-3

Monoyios, D., & Vlachos, K. (2011). Multiobjective Genetic Algorithms for Solving the Impairment-Aware Routing and Wavelength Assignment Problem. Journal of Optical Communications and Networking, 3(1), 40-47. doi: 10.1364/jocn.3.000040

Moslehi, G., & Mahnam, M. (2011). A Pareto approach to multi-objective flexible job-shop scheduling problem using particle swarm optimization and local search. International Journal of Production Economics, 129(1), 14-22. doi: 10.1016/j.ijpe.2010.08.004

Nagata, Y., Braysy, O., & Dullaert, W. (2010). A penalty-based edge assembly memetic algorithm for the vehicle routing problem with time windows. Computers & Operations Research, 37(4), 724-737. doi: 10.1016/j.cor.2009.06.022

Nascimento, M. C. V., Resende, M. G. C., & Toledo, F. M. B. (2010). GRASP heuristic with path-relinking for the multi-plant capacitated lot sizing problem. European Journal of Operational Research 200(3), 747-754. doi: 10.1016/j.ejor.2009.01.047

Nazemi, A., & Omidi, F. (2013). An efficient dynamic model for solving the shortest path problem. Transportation Research Part C-Emerging Technologies, 26, 1-19. doi: 10.1016/j.trc.2012.07.005

Ng, C. T., Kovalyov, M. Y., & Cheng, T. C. E. (2010). A simple FPTAS for a single-item capacitated economic lot-sizing problem with a monotone cost structure. European Journal of Operational Research 200(2), 621-624. doi: 10.1016/j.ejor.2009.01.040

Ngueveu, S. U., Prins, C., & Calvo, R. W. (2010). An effective memetic algorithm for the cumulative capacitated vehicle routing problem. Computers & Operations Research, 37(11), 1877-1885. doi: 10.1016/j.cor.2009.06.014

Nie, Y. (2010). A class of bush-based algorithms for the traffic assignment problem. Transportation Research Part B-Methodological, 44(1), 73-89. doi: 10.1016/j.trb.2009.06.005

Nie, Y. (2011). A cell-based Merchant-Nemhauser model for the system optimum dynamic traffic assignment problem. Transportation Research Part B-Methodological, 45(2), 329-342. doi:

10.1016/j.trb.2010.07.001

Nie, Y., & Zhang, H. M. (2010). Solving the Dynamic User Optimal Assignment Problem Considering Queue Spillback. Networks & Spatial Economics, 10(1), 49-71. doi: 10.1007/s11067-007-9022-y

Okhrin, I., & Richter, K. (2011). An O(T-3) algorithm for the capacitated lot sizing problem with minimum order quantities. European Journal of Operational Research, 211(3), 507-514. doi: 10.1016/j.ejor.2011.01.007

Ore, Oystein; Wilson, Robin J. (Revised by). Anneli Lax New Mathematical Library, Volume 34 : Graphs and Their Uses. Washington, DC, USA: Mathematical Association of America, 1990. p 33.

Ozbakir, L., Baykasoglu, A., & Tapkan, P. (2010). Bees algorithm for generalized assignment problem. Applied Mathematics and Computation, 215(11), 3782-3795. doi: 10.1016/j.amc.2009.11.018

Pan, Q. K., Suganthan, P. N., Liang, J. J., & Tasgetiren, M. F. (2011). A local-best harmony search algorithm with dynamic sub-harmony memories for lot-streaming flow shop scheduling problem. Expert Systems with Applications, 38(4), 3252-3259. doi: 10.1016/j.eswa.2010.08.111

Pan, Q. K., Tasgetiren, M. F., Suganthan, P. N., & Chua, T. J. (2011). A discrete artificial bee colony algorithm for the lot-streaming flow shop scheduling problem. Information Sciences, 181(12), 2455-2468. doi: 10.1016/j.ins.2009.12.025

Pardalos, Panos M. (Editor); Migdalas, Athanasios (Editor); Burkard, Rainer E. (Editor). Combinatorial and Global Optimization. River Edge, NJ, USA: World Scientific 2002. p 10p.

Penuel, J., Smith, J. C., & Yuan, Y. (2010). An Integer Decomposition Algorithm for Solving a Two-Stage Facility Location Problem with Second-Stage Activation Costs. Naval Research Logistics, 57(5), 391-402. doi: 10.1002/nav.20401

Perboli, G., Tadei, R., & Vigo, D. (2011). The Two-Echelon Capacitated Vehicle Routing Problem: Models and Math-Based Heuristics. Transportation Science, 45(3), 364-380. doi: 10.1287/trsc.1110.0368

Pineyro, P., & Viera, O. (2010). The economic lot-sizing problem with remanufacturing and one-way substitution. International Journal of Production Economics, 124(2), 482-488. doi: 10.1016/j.ijpe.2010.01.007

Pinto, L. L., & Pascoal, M. M. B. (2010). On algorithms for the tricriteria shortest path problem with two bottleneck objective functions. Computers & Operations Research, 37(10), 1774-1779. doi: 10.1016/j.cor.2010.01.005

Piperagkas, G. S., Konstantaras, I., Skouri, K., & Parsopoulos, K. E. (2012). Solving the stochastic dynamic lot-sizing problem through nature-inspired heuristics. Computers & Operations Research, 39(7), 1555-1565. doi: 10.1016/j.cor.2011.09.004

Pop, P. C. (2012). Generalized Network Design Problems: Modeling and Optimization. Berlin, De Gruyter.

Pop, Petrica C.; Versita (Contribution by). De Gruyter Series in Discrete Mathematics and Applications, Volume 1 : Network Design Problems : Modeling and Optimization of Generalized Network Design Problems. Hawthorne, NY, USA: Walter de Gruyter 2012. p 128.

Przybylski, A., Gandibleux, X., & Ehrgott, M. (2010). A two phase method for multi-objective integer programming and its application to the assignment problem with three objectives. Discrete Optimization, 7(3), 149-165. doi: 10.1016/j.disopt.2010.03.005

Puchinger, J., Raidl, G. R., & Pferschy, U. (2010). The Multidimensional Knapsack Problem: Structure and Algorithms. Informs Journal on Computing, 22(2), 250-265. doi: 10.1287/ijoc.1090.0344

Puerto, J., & Rodriguez-Chia, A. M. (2011). On the structure of the solution set for the single facility location problem with average distances. Mathematical Programming, 128(1-2), 373-401. doi: 10.1007/s10107-009-0308-3

Pugliese, L. D., & Guerriero, F. (2012). A computational study of solution approaches for the resource constrained elementary shortest path problem. Annals of Operations Research 201(1), 131-157. doi: 10.1007/s10479-012-1162-x

Pugliese, L. D., & Guerriero, F. (2013). Dynamic programming approaches to solve the shortest path problem with forbidden paths. Optimization Methods & Software, 28(2), 221-255. doi: 10.1080/10556788.2011.630077

Pugliese, L. D., & Guerriero, F. (2013). Shortest path problem with forbidden paths: The elementary version. European Journal of Operational Research, 227(2), 254-267. doi: 10.1016/j.ejor.2012.11.010

Ramadan, M. A., & El-Sayed, E. A. (2010). Partial eigenvalue assignment problem of high order control systems using orthogonality relations. Computers & Mathematics with Applications, 59(6), 1918-1928. doi: 10.1016/j.camwa.2009.07.063

Ramamurthy, P.. Operations Research. Daryaganj, Delhi, IND: New Age International 2007. p 214.

Ramazani, H., Shafahi, Y., & Seyedabrishami, S. E. (2010). A Shortest Path Problem in an Urban Transportation Network Based on Driver Perceived Travel Time. Scientia Iranica Transaction a-Civil Engineering, 17(4), 285-296.

Rego, C., Gamboa, D., Glover, F., & Osterman, C. (2011). Traveling salesman problem heuristics: Leading methods, implementations and latest advances. European Journal of Operational Research, 211(3), 427-441. doi: 10.1016/j.ejor.2010.09.010

Ren, Z. G., Feng, Z. R., & Zhang, A. M. (2012). Fusing ant colony optimization with Lagrangian relaxation for the multiple-choice multidimensional knapsack problem. Information Sciences, 182(1), 15-29. doi: 10.1016/j.ins.2011.07.033

Repoussis, P. P., Tarantilis, C. D., Braysy, O., & Ioannou, G. (2010). A hybrid evolution strategy for the open vehicle routing problem. Computers & Operations Research, 37(3), 443-455. doi: 10.1016/j.cor.2008.11.003

Ribas, I., Companys, R., & Tort-Martorell, X. (2011). An iterated greedy algorithm for the flowshop scheduling problem with blocking. Omega-International Journal of Management Science, 39(3), 293-301. doi: 10.1016/j.omega.2010.07.007

Ribeiro, G. M., & Laporte, G. (2012). An adaptive large neighborhood search heuristic for the cumulative capacitated vehicle routing problem. Computers & Operations Research, 39(3), 728-735. doi: 10.1016/j.cor.2011.05.005

Rondeau, Thomas W.; Bostian, Charles W.. Artificial Intelligence in Wireless Communications. Norwood, MA, USA: Artech House 2009. p 78.

Rong, A. Y., Figueira, J. R., & Klamroth, K. (2012). Dynamic programming based algorithms for the discounted {0-1} knapsack problem. Applied Mathematics and Computation, 218(12), 6921-6933. doi: 10.1016/j.amc.2011.12.068

Rong, A. Y., Figueira, J. R., & Pato, M. V. (2011). A two state reduction based dynamic programming algorithm for the bi-objective 0-1 knapsack problem. Computers & Mathematics with Applications, 62(8), 2913-2930. doi: 10.1016/j.camwa.2011.07.067

Ruiz, R., & Vazquez-Rodriguez, J. A. (2010). The hybrid flow shop scheduling problem. European Journal of Operational Research 205(1), 1-18. doi: 10.1016/j.ejor.2009.09.024

Salmasi, N., Logendran, R., & Skandari, M. R. (2010). Total flow time minimization in a flowshop sequence-dependent group scheduling problem. Computers & Operations Research, 37(1), 199-212. doi: 10.1016/j.cor.2009.04.013

Sawik, Tadeusz. Scheduling in Supply Chains Using Mixed Integer Programming. Hoboken, NJ, USA: Wiley 2011. p 29.

Sbihi, A. (2010). A cooperative local search-based algorithm for the Multiple-Scenario Max-Min Knapsack Problem. European Journal of Operational Research 202(2), 339-346. doi: 10.1016/j.ejor.2009.05.033

Schulz, T. (2011). A new Silver-Meal based heuristic for the single-item dynamic lot sizing problem with returns and remanufacturing. International Journal of Production Research, 49(9), 2519-2533. doi: 10.1080/00207543.2010.532916

Shen, Z. J. M., Zhan, R. L., & Zhang, J. W. (2011). The Reliable Facility Location Problem: Formulations, Heuristics, and Approximation Algorithms. Informs Journal on Computing, 23(3), 470-482. doi: 10.1287/ijoc.1100.0414

Shi, N. (2010). K Constrained Shortest Path Problem. Ieee Transactions on Automation Science and Engineering, 7(1), 15-23. doi: 10.1109/tase.2009.2012434

Subbu, Raj; Sanderson, Arthur C.. Network-Based Distributed Planning Using Coevolutionary Algorithms. River Edge, NJ, USA: World Scientific 2004. p 3.

Subramanian, A., Drummond, L. M. A., Bentes, C., Ochi, L. S., & Farias, R. (2010). A parallel heuristic for the Vehicle Routing Problem with Simultaneous Pickup and Delivery. Computers & Operations Research, 37(11), 1899-1911. doi: 10.1016/j.cor.2009.10.011

Sun, M. H. (2012). A tabu search heuristic procedure for the capacitated facility location problem. Journal of Heuristics, 18(1), 91-118. doi: 10.1007/s10732-011-9157-3

Szeto, W. Y., Wu, Y. Z., & Ho, S. C. (2011). An artificial bee colony algorithm for the capacitated vehicle routing problem. European Journal of Operational Research, 215(1), 126-135. doi: 10.1016/j.ejor.2011.06.006

Takaoka, T. (2013). A simplified algorithm for the all pairs shortest path problem with $O(n(2) \log n)$ expected time. Journal of

Combinatorial Optimization, 25(2), 326-337. doi: 10.1007/s10878-012-9550-3

Tang, L. X., Jiang, W., & Saharidis, G. K. D. (2013). An improved Benders decomposition algorithm for the logistics facility location problem with capacity expansions. Annals of Operations Research, 210(1), 165-190. doi: 10.1007/s10479-011-1050-9

Tarim, S. A., Dogru, M. K., Ozen, U., & Rossi, R. (2011). An efficient computational method for a stochastic dynamic lot-sizing problem under service-level constraints. European Journal of Operational Research, 215(3), 563-571. doi: 10.1016/j.ejor.2011.06.034

Tasgetiren, M. F., Suganthan, P. N., & Pan, Q. K. (2010). An ensemble of discrete differential evolution algorithms for solving the generalized traveling salesman problem. Applied Mathematics and Computation, 215(9), 3356-3368. doi: 10.1016/j.amc.2009.10.027

Tohyama, H., Ida, K., & Matsueda, J. (2011). A Genetic Algorithm for the Uncapacitated Facility Location Problem. Electronics and Communications in Japan, 94(5), 47-54. doi: 10.1002/ecj.10180

Vallada, E., & Ruiz, R. (2011). A genetic algorithm for the unrelated parallel machine scheduling problem with sequence dependent setup times. European Journal of Operational Research, 211(3), 612-622. doi: 10.1016/j.ejor.2011.01.011

Van Peteghem, V., & Vanhoucke, M. (2010). A genetic algorithm for the preemptive and non-preemptive multi-mode resource-constrained project scheduling problem. European Journal of Operational Research 201(2), 409-418. doi: 10.1016/j.ejor.2009.03.034

Viale, J. David; Carrigan, Christopher (Editor). Inventory Management : From Warehouse to Distribution Center.

Viale, J. David; Carrigan, Christopher (Editor). Inventory Management : From Warehouse to Distribution Center. Menlo Park, CA, USA: Course Technology / Cengage Learning, 1996. p 50-57.

Wang, J. B., & Guo, Q. (2010). A due-date assignment problem with learning effect and deteriorating jobs. Applied Mathematical Modelling, 34(2), 309-313. doi: 10.1016/j.apm.2009.04.020

Wang, J. B., & Wang, C. (2011). Single-machine due-window assignment problem with learning effect and deteriorating jobs. Applied Mathematical Modelling, 35(8), 4017-4022. doi: 10.1016/j.apm.2011.02.023

Wang, L., Wang, S. Y., & Xu, Y. (2012). An effective hybrid EDA-based algorithm for solving multidimensional knapsack problem. Expert Systems with Applications, 39(5), 5593-5599. doi: 10.1016/j.eswa.2011.11.058

Wang, X. J., Gao, L., Zhang, C. Y., & Shao, X. Y. (2010). A multi-objective genetic algorithm based on immune and entropy principle for flexible job-shop scheduling problem. International Journal of Advanced Manufacturing Technology, 51(5-8), 757-767. doi: 10.1007/s00170-010-2642-2

Wang, Z. C., Zhang, Y. M., Zhou, W. H., & Liu, H. F. (2012). Solving traveling salesman problem in the Adleman-Lipton model. Applied Mathematics and Computation, 219(4), 2267-2270. doi: 10.1016/j.amc.2012.08.073

Wang, Z., Du, D. L., Gabor, A. F., & Xu, D. C. (2010). An approximation algorithm for the k-level stochastic facility location problem. Operations Research Letters, 38(5), 386-389. doi: 10.1016/j.orl.2010.04.010

Wei, Z., Ge, F. Z., Lu, Y., Li, L. X., & Yang, Y. X. (2011). Chaotic ant swarm for the traveling salesman problem. Nonlinear Dynamics, 65(3), 271-281. doi: 10.1007/s11071-010-9889-x

Williams, H. Paul. Model Building in Mathematical Programming (5th Edition). Somerset, NJ, USA: Wiley 2013. p 109-220.

Wong, J. T., Su, C. T., & Wang, C. H. (2012). Stochastic dynamic lot-sizing problem using bi-level programming base on artificial intelligence techniques. Applied Mathematical Modelling, 36(5) 2003-2016. doi: 10.1016/j.apm.2011.08.017

Wu, T., Shi, L. Y., & Duffie, N. A. (2010). An HNP-MP Approach for the Capacitated Multi-Item Lot Sizing Problem With Setup Times. Ieee Transactions on Automation Science and Engineering, 7(3), 500-511. doi: 10.1109/tase.2009.2039134

Wu, T., Shi, L. Y., & Song, J. (2012). An MIP-based interval heuristic for the capacitated multi-level lot-sizing problem with setup times. Annals of Operations Research, 196(1), 635-650. doi: 10.1007/s10479-011-1026-9

Yagmahan, B., & Yenisey, M. M. (2010). A multi-objective ant colony system algorithm for flow shop scheduling problem. Expert Systems with Applications, 37(2), 1361-1368. doi: 10.1016/j.eswa.2009.06.105

Yalaoui, Alice;Chehade, Hicham;Yalaoui , Farouk;Amodeo, Lionel. (2013). Optimization of Logistics. Wiley-ISTE.

Yang, Xin-She. Introduction to Mathematical Optimization : From

Linear Programming to Metaheuristics. Cambridge, GBR: Cambridge International Science Publishing 2008. p 93.

Yang, Z., Chu, F., & Chen, H. X. (2012). A cut-and-solve based algorithm for the single-source capacitated facility location problem. European Journal of Operational Research, 221(3), 521-532. doi: 10.1016/j.ejor.2012.03.047

Yu, B., & Yang, Z. Z. (2011). An ant colony optimization model: The period vehicle routing problem with time windows. Transportation Research Part E-Logistics and Transportation Review, 47(2), 166-181. doi: 10.1016/j.tre.2010.09.010

Yu, B., Yang, Z. Z., & Xie, J. X. (2011). A parallel improved ant colony optimization for multi-depot vehicle routing problem. Journal of the Operational Research Society, 62(1), 183-188. doi: 10.1057/jors.2009.161

Zhan, Z. H., Zhang, J., Li, Y., Liu, O., Kwok, S. K., Ip, W. H., & Kaynak, O. (2010). An Efficient Ant Colony System Based on Receding Horizon Control for the Aircraft Arrival Sequencing and Scheduling Problem. Ieee Transactions on Intelligent Transportation Systems, 11(2), 399-412. doi: 10.1109/tits.2010.2044793

Zhang, G. H., Gao, L., & Shi, Y. (2011). An effective genetic algorithm for the flexible job-shop scheduling problem. Expert Systems with Applications, 38(4), 3563-3573. doi: 10.1016/j.eswa.2010.08.145

Zhang, H. Z., Beltran-Royo, C., & Constantino, M. (2010). Effective formulation reductions for the quadratic assignment problem. Computers & Operations Research, 37(11) 2007-2016. doi: 10.1016/j.cor.2010.02.001

Zhang, R., & Wu, C. (2010). A hybrid immune simulated annealing algorithm for the job shop scheduling problem. Applied Soft Computing, 10(1), 79-89. doi: 10.1016/j.asoc.2009.06.008

Zhang, X. G., Wang, Q., Chan, F. T. S., Mahadevan, S., & Deng, Y. (2014). A Physarum Polycephalum Optimization Algorithm for the Bi-objective Shortest Path Problem. International Journal of Unconventional Computing, 10(1-2), 143-162.

Zhang, Y. D., Wu, L. N., Wei, G., & Wang, S. H. (2011). A novel algorithm for all pairs shortest path problem based on matrix multiplication and pulse coupled neural network. Digital Signal Processing, 21(4), 517-521. doi: 10.1016/j.dsp.2011.02.004

Zhang, Z. H., Jiang, H., & Pan, X. Z. (2012). A Lagrangian relaxation based approach for the capacitated lot sizing problem in closed-loop supply chain. International Journal of Production Economics, 140(1), 249-255. doi: 10.1016/j.ijpe.2012.01.018

Zhao, J., Liu, Q. L., Wang, W., Wei, Z. Q., & Shi, P. (2011). A parallel immune algorithm for traveling salesman problem and its application on cold rolling scheduling. Information Sciences, 181(7), 1212-1223. doi: 10.1016/j.ins.2010.12.003

Zhao, Y. L., Zhang, J. E., Ji, Y. F., & Gu, W. Y. (2010). Routing and Wavelength Assignment Problem in PCE-Based Wavelength-Switched Optical Networks. Journal of Optical Communications and Networking, 2(4), 196-205. doi: 10.1364/jocn.2.000196

Zou, D. X., Gao, L. Q., Li, S., & Wu, J. H. (2011). Solving 0-1 knapsack problem by a novel global harmony search algorithm. Applied Soft Computing, 11(2), 1556-1564. doi: 10.1016/j.asoc.2010.07.019

Zou, D. X., Gao, L. Q., Li, S., Wu, J. H., & Wang, X. (2010). A novel global harmony search algorithm for task assignment problem. Journal of Systems and Software, 83(10), 1678-1688. doi: 10.1016/j.jss.2010.04.070

Chapter 3. Algorithms

This chapter is about algorithms used for solving optimization problems. In this chapter, the fundamentals of ant colony optimization, the cross-entropy method, the Dijkstra, the Bellman-Ford algorithm, the genetic algorithm, the Hungarian, the Jonker-Volgenant algorithm, the particle swarm, the simulated annealing method, and the Wagner-Whitin algorithm are introduced.

There exist many other numerous algorithms for solving optimization problems. This chapter only introduces few of the most popular algorithms.

3.1 Ant colony optimization

The ant colony optimization (ACO) was introduced by Marco Dorigo in the early 1990s (Sandou 2013). This algorithm is based on the social behavior of ants and allows us to solve complex optimization problems, especially integers programming problems (Sandou 2013). The ant colony optimization is a metaheuristic in which a colony of artificial ants cooperates in finding good solutions to difficult discrete optimization problems (Dorigo and Stützle 2004). Cooperation is a key element in the design of ACO algorithms: The choice is to allocate computational resources to a relatively simple set of agents (artificial ants) that communicate indirectly by stigmergy is mediated by indirect communication by the environment (Dorigo and Stützle 2004).

This algorithm is based on the social behavior of ants when they are foraging (Sandou 2013). Ant colonies, and societies of social insects in general, are systems that, despite the simplicity of their individuals, have a highly structured social organization (Dorigo and Stützle 2004) distributed. ACO metaheuristic embodies a large class of algorithms whose design is based mainly on the foraging behavior of real ants (Dehuri 2011). The route search initial random food, the concentration ranges of pheromone and follow the path of ants of the higher concentration of pheromone and pheromone is enhanced by increasing number of ants. As more and more ants follow the same path, it becomes the preferred route (Yang 2008). The amount of pheromone deposited, which may depend on the quantity and quality of the food, will guide other ants to the food source (Cho et al. 2011). Indirect communication between the ants via pheromone trails allows them to find the shortest path between their nest and food sources (Cho et al. 2011). This characteristic of real ant colonies is exploited inartificial ant colonies to solve NP-hard problems (Cho et al. 2011). For example, some preferred (often the shortest or most efficient) route emerges (Yang 2008). This is in fact a positive feedback mechanism (Yang 2008). ACO algorithms were originally designed and have a long tradition in solving a specific type of combinatorial optimization problems (ie, problems for which the construction process of the solution can be implemented by simulating a walk through a construction graph) (Dehuri 2011).

The key works using the ACO meta-heuristic has been devoted to the Travelling Salesperson Problem (TSP), a classic NP-complete problem whose main features can easily be manipulated to show the applicability of this metaheuristic (Dehuri 2011).

Ants that perform well in a given iteration influence exploration ants in future iterations. Because ants explore alternatives, the resulting pheromone trail is the result of different views on the solution space (Bonabeau et al., 1999). Even when only the best ant execution is allowed to strengthen its solution, there is a cooperative effect in time because the ants in the next iteration using the pheromone trail to guide their exploration (Bonabeau et al., 1999).

Algorithm – Outline of the ACO metaheuristic (Dehuri 2011):
1: Initialize();
2: while termination-condition is NOT TRUE do
3: BuildSolutions();
4: PheromoneUpdate();
5: DaemonActions(); // Optional
6: end while

It should be emphasized that the algorithms of ant colonies are the right tool for discrete and combinatorial optimization (Yang 2008). They have the advantages over other stochastic algorithms such as genetic algorithms and simulated annealing in the treatment of dynamic network routing problems (Yang 2008). For continuous decision variables, its performance is still being investigated (Yang 2008).

The most recent and highly cited studies about the ant colony optimization are shown in Table 3.1.

Table 3.1 Studies about the ant colony optimization

A parameter free continuous ant colony optimization algorithm for the optimal design of storm sewer networks: constrained and unconstrained approach (Afshar 2010).
A two-stage ant colony optimization algorithm to minimize the makespan on unrelated parallel machines with sequence-dependent setup times (Arnaout *et al.* 2010).
In vivo diagnosis of gastric cancer using raman endoscopy and ant colony optimization techniques (Bergholt *et al.* 2011).
Bi-objective ant colony optimization approach to optimize production and maintenance scheduling (Berrichi *et al.* 2010).
Optimizing discounted cash flows in project scheduling - An ant colony optimization approach (Chen *et al.* 2010).
A rough set approach to feature selection based on ant colony optimization (Chen *et al.* 2010).
Scheduling resource-constrained projects with ant colony optimization artificial agents (Christodoulou 2010).
Designing fuzzy-rule-based systems using continuous ant-colony optimization (Juang and Chang 2010).
An improved ant colony optimization for constrained engineering design problems (Kaveh and Talatahari 2010).
An improved ant colony optimization for the design of planar steel frames (Kaveh and Talatahari 2010).
An enhanced ant colony optimization (eaco) applied to capacitated vehicle routing problem (lee *et al.* 2010).
Integrated process planning and scheduling by an agent-based ant colony optimization (Leung *et al.* 2010).
A survey: ant colony optimization based recent research and implementation on several engineering domain (Mohan and Baskaran 2012).
Ant colony optimization algorithm to solve for the transportation problem of cross-docking network (Musa *et al.* 2010).
Ant colony optimization and the minimum spanning tree problem (Neumann and Witt 2010).
Power load forecasting using support vector machine and ant colony optimization (Niu *et al.* 2010).
A survey on parallel ant colony optimization (Pedemonte *et al.* 2011).
Comparing ant colony optimization and genetic algorithm approaches for solving traffic signal coordination under oversaturation conditions (Putha *et al.* 2012).
An improved ant colony optimization based algorithm for the capacitated arc routing problem (Santos *et al.* 2010).
Ant colony optimization for wavelet-based image interpolation using a three-component exponential mixture model (tian *et al.* 2011).
Knowledge-based ant colony optimization for flexible job shop scheduling problems (Xing *et al.* 2010).
An improved ant colony optimization algorithm for solving a complex combinatorial optimization problem (Yang and Zhuang 2010).
An ant colony optimization model: the period vehicle routing problem with time windows (Yu and Yang 2011).
A parallel improved ant colony optimization for multi-depot vehicle routing problem (Yu *et al.* 2011).
End member extraction of hyperspectral remote sensing images based on the ant colony optimization (ACO) algorithm (Zhang *et al.* 2011).

3.2 The cross-entropy (CE) method

The cross entropy method (CE) is a powerful technique to solve estimation and optimization of difficult problems, based on Kullback-Leibler (or cross-entropy) minimization (Rubinstein *et al.* 2013). It was launched in 1999 by Rubinstein, as an adaptive importance sampling to estimate the probability of rare-events (Rubinstein *et al.* 2013). Subsequent work has shown that in many optimization problems can be converted into a problem of estimation of rare events (Rubinstein *et al.* 2013). Accordingly, the method of CE can be used as randomized algorithm for optimization (Rubinstein *et al.* 2013).

3.3 The Dijkstra and the Bellman–Ford algorithm

An efficient method was developed in 1959 by the Dutch mathematician Edsger W. Dijkstra, a pioneer in the art of computer programming (Koshy 2003). Dijkstra's algorithm can find the shortest path from any vertex to any vertex in the digraph, if it exists (Koshy 2003).

An overview of Dijkstra's algorithm is described as follows (Tatipamula *et al.* 2012):

- Step 1: Set the distance for the source node to zero, and to all other nodes to infinity.
- Step 2: Mark all nodes as unvisited. Set the source node as a current.
- Step 3: For current node, consider all its unvisited neighbors and calculate their distance from the source node. If the distance is less than the distance previously recorded, the previous distance is replaced by the new one in the record. The last hop is also updated with the new leap in the previous case.
- Step 4: Select the unvisited node whose distance is shortest, and set it to a current node. Mark previous current node visited. Visited a node will not be checked over. The recorded distance of the current node is the smallest among the unvisited nodes, and it is final.
- Step 5: If all nodes have been visited, the process is complete. If not, repeat step 3.

The Bellman-Ford: Before describing the algorithm in detail, a few preliminaries are necessary (Benvenuto and Zorzi 2011). The Bellman-Ford algorithm derives its solution iteratively by building sequences of nodes which are temporary best walks (Benvenuto Zorzi and 2011). We say "walks" because we do not necessarily advocate visit the same node more than once. Thus, the algorithm seems at first to find walks, not paths (Benvenuto and Zorzi 2011). Recall that a path is always one walk (and thus a shortest path is an optimal walk), but not vice versa; it depends if walking contains cycles or not (only in the latter case is the walk path also a path) (Benvenuto and Zorzi 2011).

Dijkstra's algorithm: Dijkstra's algorithm is similar to the Bellman-Ford algorithm, although it has a computational complexity of lower the worst-case (Benvenuto and Zorzi 2011).

Dijkstra's algorithm uses an iterative approach, the evaluation of path costs and update (Benvenuto and Zorzi 2011). Ford algorithm Bellman-blindly reiterated all the temporary costs, and recognized that there was no need for further updates only after an iteration that left unchanged all costs (Benvenuto and Zorzi 2011). Dijkstra's algorithm takes a slightly different approach: it determines the shortest to an order of increasing cost paths (Benvenuto and Zorzi 2011).

3.4 The genetic algorithm (GA)

The genetic algorithm (GA), developed by John Holland and his colleagues in the 1960s and 1970s, is a model or abstraction of biological evolution based on Charles Darwin's theory of natural selection (Yang 2010). Holland was the first to use the crossover and recombination, mutation and selection in the study of adaptive and artificial systems (Yang 2010).

GA encodes the decision variables or input parameters of the problem solution strings of finite length (Rao and Savsani 2012). Although the traditional optimization techniques work directly with decision variables or input parameters, genetic algorithms usually work with coding. Genetic algorithms start looking in a population of encoded solutions instead of a single point in the solution space (Rao and Savsani 2012). The initial population of individuals is created randomly. Genetic algorithms use genetic operators to create global optimal solutions based on the solutions to the current population (Rao and Savsani 2012). The most popular genetic operators are (1) selection, (2) crossover and (3) mutation (Rao and Savsani 2012). The newly generated individuals replace the old population, and the product of evolutionary processes until some termination criteria are satisfied (Rao and Savsani 2012). These genetic operators form the core of the genetic algorithm as a strategy for problem solving (Yang 2010). Since then, many variations of genetic algorithms have been developed and applied to a wide range of optimization problems, graph coloring pattern recognition, discrete systems such as the traveling salesman problem for continuous systems such the effective wing design in aerospace engineering, and financial market multiobjective engineering optimization (Yang 2010).

There are many advantages of genetic algorithms optimization algorithms, and two most notable benefits include: the ability to cope with complex problems of optimization and parallelism (Yang 2010). Genetic algorithms can cope with different types of optimization if the objective function (fitness) is stationary or non-stationary (changing over time), linear or nonlinear, continuous or discontinuous, or random noise (Yang 2010). As multiple offsprings in an act of the population as independent agents, population (or subgroup) can explore the search space in several directions simultaneously (Yang 2010). This feature makes it ideal for parallel algorithms implementation (Yang 2010). Different parameters and even different groups of encoded strings can be handled simultaneously (Yang 2010).

However, genetic algorithms also have some disadvantages (Yang 2010). The formulation of the fitness function, using the population size, the choice of important parameters such as the rate of mutation and crossover, and the criteria for selection of the new population should be carefully conducted (Yang 2010). Any wrong choice, it will be difficult for the algorithm to converge, or it simply produces impossible results (Yang 2010). Despite these disadvantages, genetic algorithms are one of the optimization algorithms widely used in modern nonlinear optimization (Yang 2010).

This is often done by the following procedure (Yang 2010):

- Encoding objectives or optimization
- The initialization of a population of individuals; functions; Defining a fitness function or selection criterion;
- The evaluation of the fitness of all individuals in the population;
- Creation of a new population by crossing performance, and mutation, fitness proportional reproduction etc;
- Changes in population until some stopping criteria are met;
- Decoding results for the solution to this problem.

The most recent and highly cited studies about the genetic algorithm optimization are shown in Table 3.2.

Table 3.2 Studies about the optimization by using genetic algorithm

Exergoenvironmental analysis and optimization of a cogeneration plant system using multimodal genetic algorithm (mga) (Ahmadi and Dincer 2010).
Application of pso (particle swarm optimization) and ga (genetic algorithm) techniques on demand estimation of oil in Iran (Assareh et al. 2010).
A comparison of feature selection models utilizing binary particle swarm optimization and genetic algorithm in determining coronary artery disease using support vector machine (Babaoglu et al. 2010).
Exergoeconomic analysis and optimization of an integrated solar combined cycle system (isccs) using genetic algorithm (Baghernejad and Yaghoubi 2011).
Rsm and ann modeling for electrocoagulation of copper from simulated wastewater: multi objective optimization using genetic algorithm approach (Bhatti et al. 2011).
Tracing sediment loss from eroding farm tracks using a geochemical fingerprinting procedure combining local and genetic algorithm optimisation (Collins et al. 2010).
Damping of power system oscillations using genetic algorithm and particle swarm optimization (Eslami et al. 2010).
Genetic algorithm optimization in drug design qsar: bayesian-regularized genetic neural networks (brgnn) and genetic algorithm-optimized support vectors machines (ga-svm) (Fernandez et al. 2011).
Mathematical modeling and genetic algorithm optimization of clove oil extraction with supercritical carbon dioxide (Hatami et al. 2010).
Modeling and multi-objective exergy based optimization of a combined cycle power plant using a genetic algorithm (Kaviri et al. 2012).
A hybrid of genetic algorithm and particle swarm optimization for solving bi-level linear programming problem - a case study on supply chain model (Kuo and Han 2011).
Multiobjective optimization of building design using trnsys simulations, genetic algorithm, and artificial neural network (Magnier and Haghighat 2010).
A combination of genetic algorithm and particle swarm optimization for optimal dg location and sizing in distribution systems (Moradi and Abedini 2012).
Optimization of an artificial neural network topology using coupled response surface methodology and genetic algorithm for fluidized bed drying (Nazghelichi et al. 2011).
Comparing ant colony optimization and genetic algorithm approaches for solving traffic signal coordination under oversaturation conditions (Putha et al. 2012).
A genetic-algorithm-aided stochastic optimization model for regional air quality management under uncertainty (Qin et al. 2010).
Thermal-economic multi-objective optimization of plate fin heat exchanger using genetic algorithm (Sanaye and Hajabdollahi 2010).
A case study on optimization of biomass flow during single-screw extrusion cooking using genetic algorithm (ga) and response surface method (rsm) (Shankar et al. 2010).
Simultaneous optimization of luminance and color chromaticity of phosphors using a nondominated sorting genetic algorithm (Sharma et al. 2010).
Artificial neural network modeling and genetic algorithm based medium optimization for the improved production of marine biosurfactant (Sivapathasekaran et al. 2010).
Parametric optimization design for supercritical co2 power cycle using genetic algorithm and artificial neural network (Wang et al. 2010).
Optimization of capacity and operation for cchp system by genetic algorithm (Wang et al. 2010).
Accounting for greenhouse gas emissions in multiobjective genetic algorithm optimization of water distribution systems (Wu et al. 2010).
An adaptive reanalysis method for genetic algorithm with application to fast truss optimization (Xu et al. 2010).
Artificial neural network-genetic algorithm based optimization for the immobilization of cellulase on the smart polymer eudragit l-100 (Zhang et al. 2010).

3.5 The Hungarian algorithm & the Jonker-Volgenant algorithm

The algorithm is defined here for minimizing the total cost of assignment. The agents form the rows of the matrix and the tasks form the columns with the entry c ij being the cost of using agent i to perform task j. If the matrix is not square, we make it square by adding row(s) or column(s) of zeroes when necessary. A maximum assignment can be converted into a minimum assignment by replacing each entry c ij with C − c ij , where C is the maximum value in the assignment matrix. The Hungarian algorithm proceeds in the following steps:

1. Subtract the minimum number in each row from each entry in the entire row.
2. Subtract the minimum number in each column from each entry in the entire column.
3. Cover all zeroes in the matrix with as few lines (horizontal and/or vertical only) as possible. Let k be the number of lines and n the size of the matrix.
   - If k < n, let m be the minimum uncovered number. Subtract m from every uncovered number and add m to every number covered by two lines. Go back to step 3.
   - If k = n, go to step 4.
4. Starting with the top row, work your way downwards as you make assignments. An assignment can be (uniquely) made when there is exactly one zero in a row. Once an assignment is made, delete that row and column from the matrix.

The linear assignment problem (LAP) is useful as a relaxation for difficult combinatorial optimization problems like quadratic assignment, and traveling salesman. Furthermore, theoretical developments for the LAP can often be extended to other problems, such as minimum cost flow and transportation.

The computational results show that the average computation times of the algorithm LAPJV are uniformly lower than the best of other algorithms. The code is of moderate size, and the memory requirements are small. The algorithm is suited for both dense and sparse assignment problems, and its sensitivity to cost range is relatively low.

3.6 The particle swarm optimization

Particle swarm optimization (PSO) was developed by Kennedy and Eberhart in 1995 based on the behavior of swarms such as fish schooling and birds in nature (Yang 2008). Many algorithms (such as ant colony algorithms and virtual ants algorithms) use the behavior of the so-called swarm intelligence (Yang 2008). Although particle swarm optimization has many similarities with genetic algorithms and virtual ants algorithms, but it is much simpler because it does not use mutation / crossover operators or pheromone (Yang 2008). Swarm intelligence with the collective behavior of systems with many interacting locally with one another and with their environment offers, and the ways of using decentralized control and self-organization to achieve their goals (Cho *et al.* 2011). In computing, particle swarm optimization (PSO) is a calculation method that optimizes a problem by iteratively trying to improve a candidate solution with regard to a given measure of quality (Garrett 2012). These methods are commonly known as metaheuristics as they make few or no assumptions about the problem optimized and can search very large spaces of candidate solutions (Garrett 2012). However, metaheuristics such as PSO do not guarantee an optimal solution is ever found (Garrett 2012).

PSO is a search algorithm based on population-based simulation of the social behavior of birds within a flock (Cho *et al.* 2011). Originally, it was adopted for the formation of neural networks and optimization of nonlinear function, and has quickly become a popular global optimizer, especially in problems where the decision variables are real numbers (Cho *et al.* 2011). This algorithm search space of an objective function by adjusting the trajectories of individual agents, called particles, such as the path of pieces formed by position vectors in a quasi-random manner (Yang 2008). Particle motion has two main components: a stochastic component and a deterministic component (Yang 2008). The particle is attracted towards the position of the current global best while at the same time; they tend to move randomly (Yang 2008). When a particle is a place that is better than any of the locations previously found, it updates as the best new current of the particle i. It is better aware of all the particles n (Yang 2008). The goal is to find the best in all the world's best current until the goal is not to improve or after a certain number of iterations (Yang 2008).

More specifically, does not use the PSO gradient problem being optimized, which means does not need PSO to the optimization problem differentiable necessary by conventional methods such as optimization of the gradient descent methods and equivalents Newton (Garrett 2012). PSO can also be used on optimization problems that are partially irregular, noisy, change over time, etc. (Garrett 2012). PSO optimizes a problem by having a population of candidate solutions, here dubbed particles, and moving the particles around the search space by simple mathematical formulas (Garrett 2012). The movements of the particles are guided by the most found in the search space positions are updated as better positions are found by particles (Garrett 2012).

PSO has become so popular because its main algorithm is relatively simple and easy to implement (Cho *et al.* 2011). It is also simple and has proven to be very effective in a wide variety of applications with very good results at a very low computational cost (Cho *et al.* 2011).

Particle Swarm Optimization Algorithm (Cho *et al.* 2011)
(1) BEGIN
(2) Parameter settings and initialization of swarm.
(3) Evaluate fitness and locate the leader (i.e., initialize p best and g best).
(4) I = 0 /* I = Iteration count */
(5) WHILE (the stopping criterion is not met, say, I < Imax)
(6) DO
(7) FOR each particle
(8) Update position & velocity (flight) as per equations
(9) Evaluate fitness

(10) Update p best
(11) END FOR
(12) Update leader (i.e.,g best)
(13) I++
(14) END WHILE
(15) END

It has the attributes common evolutionary computation, including the initialization with a population of random solutions and the search for an optimum by updating generations (Rao and Savsani 2012). Possible solutions, called particles, are then transported through the problem space by following the current optimum particles (Rao and Savsani 2012). The concept of particle swarm was originally a simulation of a simplified social system (Rao and Savsani 2012). The original intent was to graphically simulate the graceful but unpredictable choreography of a flock of birds. Each particle keeps track of its coordinates in the problem space, which are associated with the best solution (fitness) it has achieved so far (Rao and Savsani 2012).

The most recent and highly cited studies about the particle swarm optimization are shown in Table 3.3.

Table 3.3 Studies about the particle swarm optimization

A novel set-based particle swarm optimization method for discrete optimization problems (Chen *et al.* 2010).
Gaussian quantum-behaved particle swarm optimization approaches for constrained engineering design problems (Coelho 2010).
An improved particle swarm optimization (pso)-based mppt for pv with reduced steady-state oscillation (Ishaque *et al.* 2012).
Niching without niching parameters: particle swarm optimization using a ring topology (Li 2010).
An automatic group composition system for composing collaborative learning groups using enhanced particle swarm optimization (Lin *et al.* 2010).
Hybridizing particle swarm optimization with differential evolution for constrained numerical and engineering optimization (Liu *et al.* 2010).
Quantum-inspired particle swarm optimization for valve-point economic load dispatch (Meng *et al.* 2010).
A combination of genetic algorithm and particle swarm optimization for optimal dg location and sizing in distribution systems (Moradi and Abedini 2012).
A pareto approach to multi-objective flexible job-shop scheduling problem using particle swarm optimization and local search (Moslehi and Mahnam 2011).
A novel particle swarm optimization algorithm with adaptive inertia weight (Nickabadi *et al.* 2011).
A new fuzzy adaptive hybrid particle swarm optimization algorithm for non-linear, non-smooth and non-convex economic dispatch problem (Niknam 2010).
A new fuzzy adaptive particle swarm optimization for non-smooth economic dispatch (Niknam *et al.* 2010).
An improved particle swarm optimization for nonconvex economic dispatch problems (Park *et al.* 2010).
Simplifying particle swarm optimization (Pedersen and Chipperfield 2010).
Handling sideband radiations in time-modulated arrays through particle swarm optimization (Poli *et al.* 2010).
Cellular particle swarm optimization (Shi *et al.* 2011).
A discrete particle swarm optimization method for feature selection in binary classification problems (Unler and Murat 2010).
An improved evolutionary method with fuzzy logic for combining particle swarm optimization and genetic algorithms (Valdez *et al.* 2011).
Real-time pid control strategy for maglev transportation system via particle swarm optimization (Wai *et al.* 2011).
Particle swarm optimization for redundant building cooling heating and power system (Wang *et al.* 2010).
Self-adaptive learning based particle swarm optimization (Wang *et al.* 2011).
Crystal structure prediction via particle-swarm optimization (Wang *et al.* 2010).
Prediction of parkinson's disease tremor onset using a radial basis function neural network based on particle swarm optimization (Wu *et al.* 2010).
A hybrid-forecasting model based on gaussian support vector machine and chaotic particle swarm optimization (Wu 2010).
Orthogonal learning particle swarm optimization (Zhan *et al.* 2011).

3.6.1 Snapshots of examples solved by using PSO

The Himmelblau test function

The Himmelblau test function is defined as

$$F(x1, x2) = (x1^2 + x2 - 11)^2 + (x1 + x2^2 - 7)^2$$

This function is depicted in Figure 3.1 below.

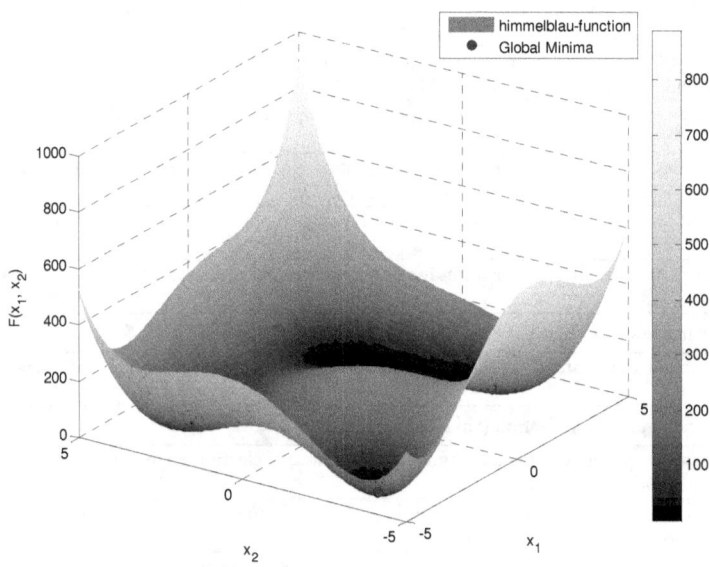

Figure 3.1 The Himmelblau test function

Optimal solution for the Himmelblau test function is obtained by PSO. See Figure 3.2.

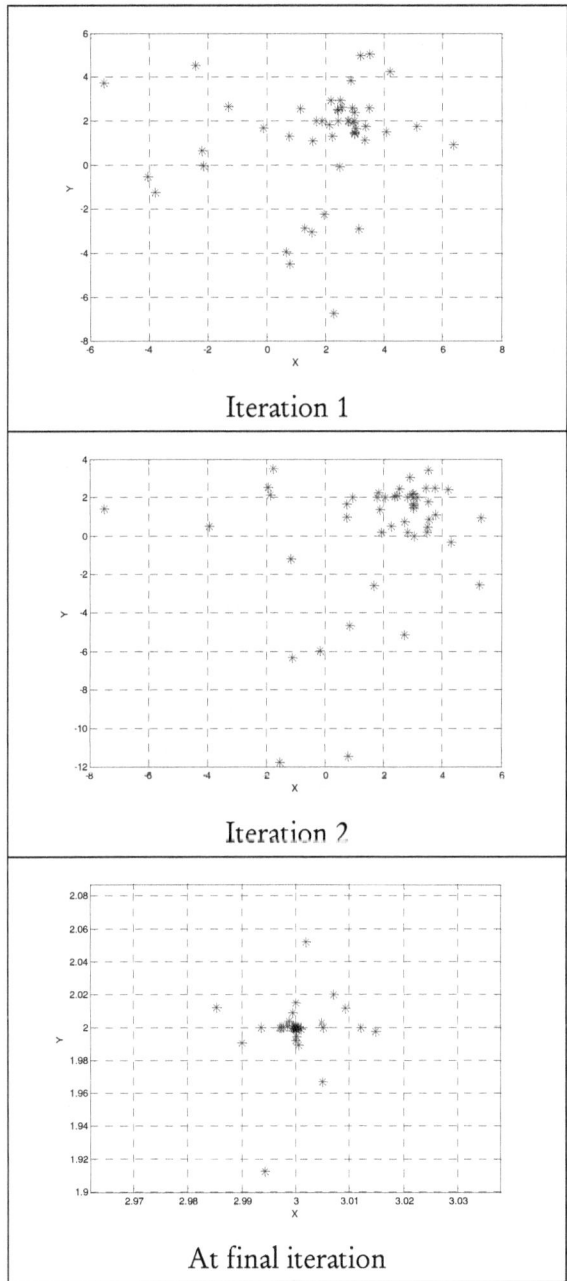

Figure 3.2 PSO Iterations

The "pen holder" test function

The "pen holder" test function is defined as

$$F(x1, x2) = -e^{\left(-\left|\frac{1}{\cos(x1)\cos(x2)\,e^{\left|1-\frac{\sqrt{x1^2+x2^2}}{\pi}\right|}}\right|\right)}$$

The function is depicted in Figure 3.3 below.

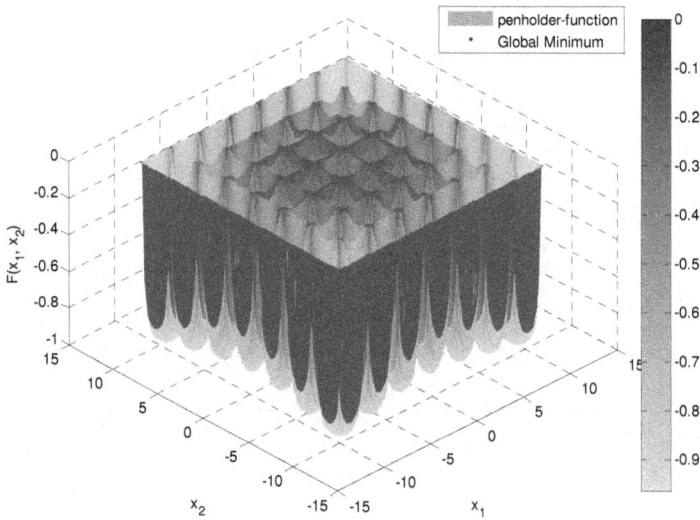

Figure 3.3 The "pen holder" test function

Optimal solution for the "pen holder" test function is obtained by PSO. See Figure 3.4.

Optimization of Logistics: Theory and Practice

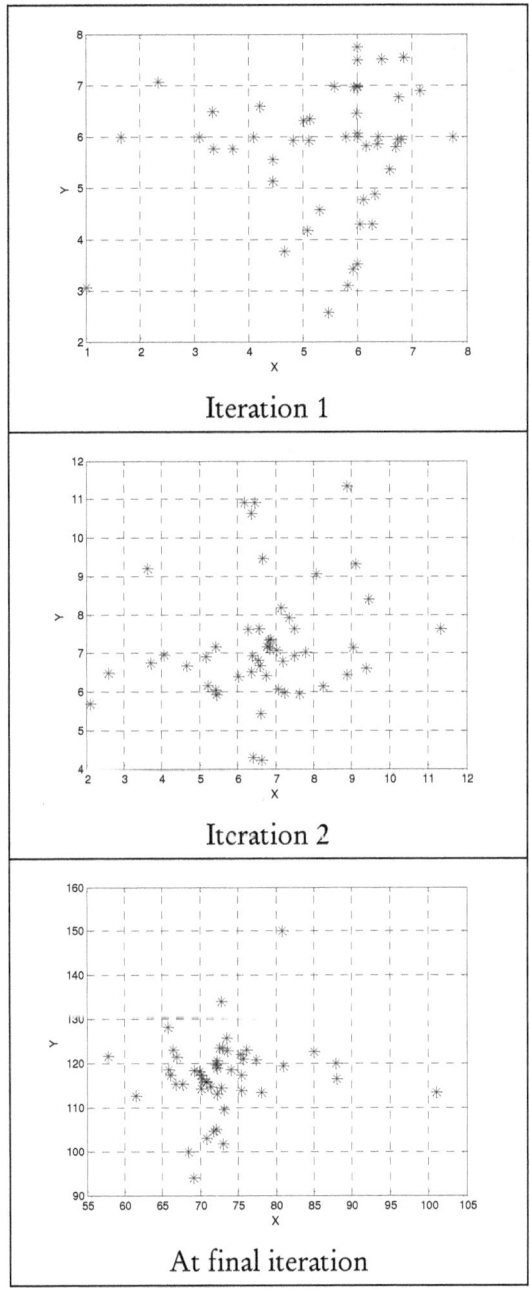

Figure 3.4 PSO Iterations

3.7 The simulated annealing method

Simulated annealing (SA) is a random search technique for global optimization problems, and it mimics the annealing process in the treatment of materials where the metal cools and solidifies in a crystalline state than the minimum energy and a larger to reduce crystal defects in metallic structures (Yang 2008). The method comprises annealing the precise control of the temperature and cooling rate (often called annealing schedule) (Yang 2008).

Simulated annealing (SA) is a generic probabilistic metaheuristic for the global optimization problem of applied mathematics, namely locating a good approximation to the global optimum of a given in a large search space (Garrett 2012) function. It is often used when the search space is discrete (eg, all tours that visit a set of cities) (Garrett 2012).

For some problems, simulated annealing may be more effective than exhaustive enumeration provided that the goal is simply to find a good solution in an acceptable period of time, rather than the best possible solution (Garrett 2012).

The most recent and highly cited studies about the simulated annealing optimization are shown in Table 3.4.

Table 3.4 Studies about the optimization by using simulated annealing

Prediction of principal ground-motion parameters using a hybrid method coupling artificial neural networks and simulated annealing (Alavi and Gandomi 2011).
Robust optimization with simulated annealing (Bertsimas and Nohadani 2010).
Multi-objective optimization of a stochastic assembly line balancing: a hybrid simulated annealing algorithm (Cakir *et al.* 2011).
Optimization of wire electrical discharge machining for pure tungsten using a neural network integrated simulated annealing approach (Chen *et al.* 2010).
Solving the traveling salesman problem based on the genetic simulated annealing ant colony system with particle swarm optimization techniques (Chen and Chien 2011).
Size optimization of a pv/wind hybrid energy conversion system with battery storage using simulated annealing (Ekren and Ekren 2010).
Solving the traveling salesman problem based on an adaptive simulated annealing algorithm with greedy search (Geng *et al.* 2011).
Solving a single-machine scheduling problem with maintenance, job deterioration and learning effect by simulated annealing (Ghodratnama *et al.* 2010).
Traffic flow forecasting by seasonal svr with chaotic simulated annealing algorithm (Hong 2011).
Simulated annealing for optimal ship routing (Kosmas and Vlachos 2012).
Using simulated annealing to minimize fuel consumption for the time-dependent vehicle routing problem (Kuo 2010).
Branch-and-bound and simulated annealing algorithms for a two-agent scheduling problem (Lee *et al.* 2010).
Electromagnetism-like mechanism and simulated annealing algorithms for flowshop scheduling problems minimizing the total weighted tardiness and makespan (Naderi *et al.* 2010).
Balancing stochastic two-sided assembly lines: a chance-constrained, piecewise-linear, mixed integer program and a simulated annealing algorithm (Ozcan 2010).
Reverse logistics network design using simulated annealing (Pishvaee *et al.* 2010).
Fuzzy control systems with reduced parametric sensitivity based on simulated annealing (Precup *et al.* 2012).
3d face recognition using simulated annealing and the surface interpenetration measure (Queirolo *et al.* 2010).
C-psa: constrained pareto simulated annealing for constrained multi-objective optimization (Singh *et al.* 2010).
Intelligent energy resource management considering vehicle-to-grid: a simulated annealing approach (Sousa *et al.* 2012).
Fast and accurate protein substructure searching with simulated annealing and gpus (Stivala *et al.* 2010).
A monte carlo/simulated annealing algorithm for sequential resonance assignment in solid state nmr of uniformly labeled proteins with magic-angle spinning (Tycko and Hu 2010).
Coupled simulated annealing (Xavier-de-Souza *et al.* 2010).
A simulated annealing heuristic for the capacitated location routing problem (Yu *et al.* 2010).
A hybrid immune simulated annealing algorithm for the job shop scheduling problem (Zhang and Wu 2010).
A simulated annealing algorithm based on block properties for the job shop scheduling problem with total weighted tardiness objective (Zhang and Wu 2011).

3.7.1 Snapshots of some examples solved by using simulated annealing method

The "test tube holder" sample test function

The "test tube holder" sample test function is defined as

$$F(x1, x2) = -4 \left| \sin(x1) \cos(x2) e^{\left| \cos\left(\frac{x1^2}{200} + \frac{x2^2}{200}\right) \right|} \right|$$

This function is depicted in Figure 3.5 below.

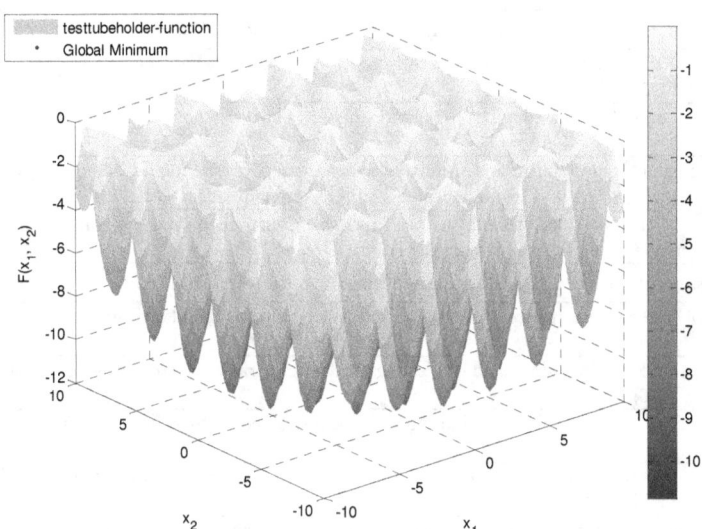

Figure 3.5 The "test tube holder" test function

The solution parameters are shown below.
Initial temperature   : 1
Final temperature   : 2.5711e-007
Consecutive rejections : 1239
Number of function calls : 6836
Total final loss   : -10.8723
x   =   1.5707 -0.0000
f   =   -10.8723

The Levi13 sample test function

The Levi13 test function is defined as

$$F(x1, x2) = \sin(3\pi x1)^2 + (x1 - 1)^2 (1 + \sin(3\pi x2)^2) + (x2 - 1)^2 (1 + \sin(2\pi x2)^2)$$

The function is depicted in Figure 3.6 below.

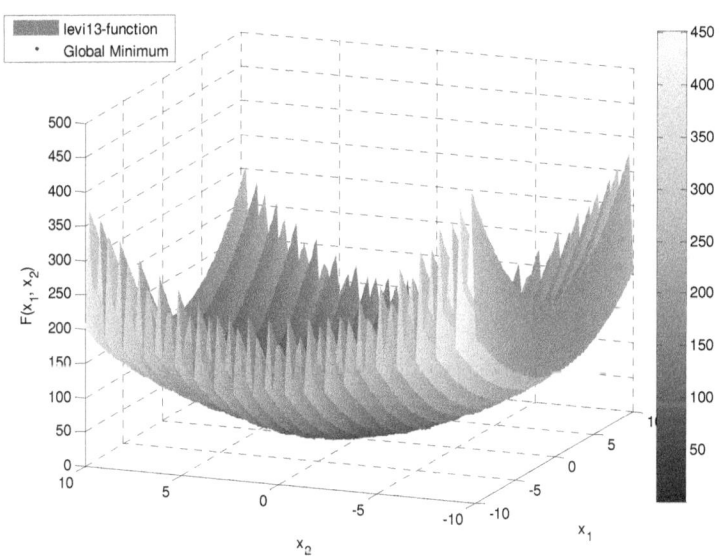

Figure 3.6 The Levi13 test function

The solution parameters by using SA are shown below.

Initial temperature   : 1
Final temperature   : 5.02168e-007
Consecutive rejections   : 1085
Number of function calls : 9342
Total final loss   : 1.98092

x =   0.0113   0.0079
f =   1.9809

3.8 The Wagner-Whitin algorithm

Wagner and Whitin studied the uncapacitated model with set-up of fixed costs and linear inventory and production costs (Woodruff 2002). Their main contribution was to demonstrate that an optimal replenishment policy is one in which production is achieved when inventory is zero (Woodruff 2002). In addition, they proposed an algorithm for dynamic programming before effective to solve the problem (Woodruff 2002).

Wagner and Wagner Whitin Whitin and (1958) have shown that the algorithm optimal solution in any period is either (Wee 2011):

1) the initial stock is zero and lot size is positive,
2) or the initial stock is positive and the lot size is zero,
3) or initial inventory both lot-size are zero

This property is known as the Wagner and Whitin property (Wee 2011). All heuristic procedures listed above are lot size for each period that has property W-W (Wee 2011). This implies that the lot size in any period must be exactly the requirements of the proceeding several periods (Wee 2011).

References

Afshar, M. H. (2010). A parameter free Continuous Ant Colony Optimization Algorithm for the optimal design of storm sewer networks: Constrained and unconstrained approach. Advances in Engineering Software, 41(2), 188-195. doi: 10.1016/j.advengsoft.2009.09.009

Ahmadi, P., & Dincer, I. (2010). Exergoenvironmental analysis and optimization of a cogeneration plant system using Multimodal Genetic Algorithm (MGA). Energy, 35(12), 5161-5172. doi: 10.1016/j.energy.2010.07.050

Alavi, A. H., & Gandomi, A. H. (2011). Prediction of principal ground-motion parameters using a hybrid method coupling artificial neural networks and simulated annealing. Computers & Structures, 89(23-24), 2176-2194. doi: 10.1016/j.compstruc.2011.08.019

Arnaout, J. P., Rabadi, G., & Musa, R. (2010). A two-stage Ant Colony Optimization algorithm to minimize the makespan on unrelated parallel machines with sequence-dependent setup times. Journal of Intelligent Manufacturing, 21(6), 693-701. doi: 10.1007/s10845-009-0246-1

Assareh, E., Behrang, M. A., Assari, M. R., & Ghanbarzadeh, A. (2010). Application of PSO (particle swarm optimization) and GA (genetic algorithm) techniques on demand estimation of oil in Iran. Energy, 35(12), 5223-5229. doi: 10.1016/j.energy.2010.07.043

Babaoglu, I., Findik, O., & Ulker, E. (2010). A comparison of feature selection models utilizing binary particle swarm optimization and genetic algorithm in determining coronary artery disease using support vector machine. Expert Systems with Applications, 37(4), 3177-3183. doi: 10.1016/j.eswa.2009.09.064

Baghernejad, A., & Yaghoubi, M. (2011). Exergoeconomic analysis and optimization of an Integrated Solar Combined Cycle System (ISCCS) using genetic algorithm. Energy Conversion and Management, 52(5), 2193-2203. doi: 10.1016/j.enconman.2010.12.019

Benvenuto, Nevio (Editor); Zorzi, Michele (Editor). Principles of Communications Networks and Systems. Hoboken, NJ, USA: Wiley 2011. p 723.

Benvenuto, Nevio (Editor); Zorzi, Michele (Editor). Principles of

Communications Networks and Systems. Hoboken, NJ, USA: Wiley 2011. p 726.

Benvenuto, Nevio (Editor); Zorzi, Michele (Editor). Principles of Communications Networks and Systems. Hoboken, NJ, USA: Wiley 2011. p 727.

Bergholt, M. S., Zheng, W., Lin, K., Ho, K. Y., Teh, M., Yeoh, K. G., . . . Huang, Z. W. (2011). In vivo diagnosis of gastric cancer using Raman endoscopy and ant colony optimization techniques. International Journal of Cancer, 128(11), 2673-2680. doi: 10.1002/ijc.25618

Berrichi, A., Yalaoui, F., Amodeo, L., & Mezghiche, M. (2010). Bi-Objective Ant Colony Optimization approach to optimize production and maintenance scheduling. Computers & Operations Research, 37(9), 1584-1596. doi: 10.1016/j.cor.2009.11.017

Bertsimas, D., & Nohadani, O. (2010). Robust optimization with simulated annealing. Journal of Global Optimization, 48(2), 323-334. doi: 10.1007/s10898-009-9496-x

Bhatti, M. S., Kapoor, D., Kalia, R. K., Reddy, A. S., & Thukral, A. K. (2011). RSM and ANN modeling for electrocoagulation of copper from simulated wastewater: Multi objective optimization using genetic algorithm approach. Desalination, 274(1-3), 74-80. doi: 10.1016/j.desal.2011.01.083

Bonabeau, Eric; Theraulaz, Guy; Dorigo, Marco. Swarm Intelligence: From Natural to Artificial Systems. Cary, NC, USA: Oxford University Press, 1999. p 54.

Bonabeau, Eric; Theraulaz, Guy; Dorigo, Marco. Swarm Intelligence: From Natural to Artificial Systems. Cary, NC, USA: Oxford University Press, 1999. p 54.

Cakir, B., Altiparmak, F., & Dengiz, B. (2011). Multi-objective optimization of a stochastic assembly line balancing: A hybrid simulated annealing algorithm. Computers & Industrial Engineering, 60(3), 376-384. doi: 10.1016/j.cie.2010.08.013

Chen, H. C., Lin, J. C., Yang, Y. K., & Tsai, C. H. (2010). Optimization of wire electrical discharge machining for pure tungsten using a neural network integrated simulated annealing approach. Expert Systems with Applications, 37(10), 7147-7153. doi: 10.1016/j.eswa.2010.04.020

Chen, S. M., & Chien, C. Y. (2011). Solving the traveling salesman problem based on the genetic simulated annealing ant colony system with particle swarm optimization techniques. Expert

Systems with Applications, 38(12), 14439-14450. doi: 10.1016/j.eswa.2011.04.163

Chen, W. N., Zhang, J., Chung, H. S. H., Huang, R. Z., & Liu, O. (2010). Optimizing Discounted Cash Flows in Project Scheduling-An Ant Colony Optimization Approach. Ieee Transactions on Systems Man and Cybernetics Part C-Applications and Reviews, 40(1), 64-77. doi: 10.1109/tsmcc.2009.2027335

Chen, W. N., Zhang, J., Chung, H. S. H., Zhong, W. L., Wu, W. G., & Shi, Y. H. (2010). A Novel Set-Based Particle Swarm Optimization Method for Discrete Optimization Problems. Ieee Transactions on Evolutionary Computation, 14(2), 278-300. doi: 10.1109/tevc.2009.2030331

Chen, Y. M., Miao, D. Q., & Wang, R. Z. (2010). A rough set approach to feature selection based on ant colony optimization. Pattern Recognition Letters, 31(3), 226-233. doi: 10.1016/j.patrec.2009.10.013

Cho, S.-B., et al. (2011). Integration of Swarm Intelligence and Artificial Neural Network. New Jersey, World Scientific.

Cho, S.-B., et al. (2011). Integration of Swarm Intelligence and Artificial Neural Network. New Jersey, World Scientific.

Cho, S.-B., et al. (2011). Integration of Swarm Intelligence and Artificial Neural Network. New Jersey, World Scientific.

Christodoulou, S. M. (2010). Scheduling Resource-Constrained Projects with Ant Colony Optimization Artificial Agents. Journal of Computing in Civil Engineering, 24(1), 45-55. doi: 10.1061/(asce)0887-3801(2010)24:1(45)

Coelho, L. D. (2010). Gaussian quantum-behaved particle swarm optimization approaches for constrained engineering design problems. Expert Systems with Applications, 37(2), 1676-1683. doi: 10.1016/j.eswa.2009.06.044

Collins, A. L., Zhang, Y., Walling, D. E., Grenfell, S. E., & Smith, P. (2010). Tracing sediment loss from eroding farm tracks using a geochemical fingerprinting procedure combining local and genetic algorithm optimisation. Science of the Total Environment, 408(22), 5461-5471. doi: 10.1016/j.scitotenv.2010.07.066

Dehuri, Satchidananda. Integration of Swarm Intelligence and Artificial Neural Network. SGP: World Scientific Publishing Co. 2011. p 69.

Dehuri, Satchidananda. Integration of Swarm Intelligence and Artificial Neural Network. SGP: World Scientific Publishing Co.

2011. p 69.
Dorigo, Marco; Stützle, Thomas. Ant Colony Optimization. Cambridge, MA, USA: MIT Press 2004. p 1.
Dorigo, Marco; Stützle, Thomas. Ant Colony Optimization. Cambridge, MA, USA: MIT Press 2004. p 33.
Ekren, O., & Ekren, B. Y. (2010). Size optimization of a PV/wind hybrid energy conversion system with battery storage using simulated annealing. Applied Energy, 87(2), 592-598. doi: 10.1016/j.apenergy.2009.05.022
Eslami, M., Shareef, H., Mohamed, A., & Khajehzadeh, M. (2010). Damping of Power System Oscillations Using Genetic Algorithm and Particle Swarm Optimization. International Review of Electrical Engineering-Iree, 5(6), 2745-2753.
Fernandez, M., Caballero, J., Fernandez, L., & Sarai, A. (2011). Genetic algorithm optimization in drug design QSAR: Bayesian-regularized genetic neural networks (BRGNN) and genetic algorithm-optimized support vectors machines (GA-SVM). Molecular Diversity, 15(1), 269-289. doi: 10.1007/s11030-010-9234-9
Garrett, Yevette. (2012). Optimization Algorithms Handbook. University Publications.
Garrett, Yevette. (2012). Optimization Algorithms Handbook. University Publications.
Geng, X. T., Chen, Z. H., Yang, W., Shi, D. Q., & Zhao, K. (2011). Solving the traveling salesman problem based on an adaptive simulated annealing algorithm with greedy search. Applied Soft Computing, 11(4), 3680-3689. doi: 10.1016/j.asoc.2011.01.039
Ghodratnama, A., Rabbani, M., Tavakkoli-Moghaddam, R., & Baboli, A. (2010). Solving a single-machine scheduling problem with maintenance, job deterioration and learning effect by simulated annealing. Journal of Manufacturing Systems, 29(1), 1-9. doi: 10.1016/j.jmsy.2010.06.004
Hatami, T., Meireles, M. A. A., & Zahedi, G. (2010). Mathematical modeling and genetic algorithm optimization of clove oil extraction with supercritical carbon dioxide. Journal of Supercritical Fluids, 51(3), 331-338. doi: 10.1016/j.supflu.2009.10.001
Hong, W. C. (2011). Traffic flow forecasting by seasonal SVR with chaotic simulated annealing algorithm. Neurocomputing, 74(12-13) 2096-2107. doi: 10.1016/j.neucom.2010.12.032
Ibe, Oliver C.. Fundamentals of Stochastic Networks. Hoboken, NJ, USA: Wiley 2011. p 184.

Ishaque, K., Salam, Z., Amjad, M., & Mekhilef, S. (2012). An Improved Particle Swarm Optimization (PSO)-Based MPPT for PV With Reduced Steady-State Oscillation. Ieee Transactions on Power Electronics, 27(8), 3627-3638. doi: 10.1109/tpel.2012.2185713

Jonker, R. and A. Volgenant (1987). "A shortest augmenting path algorithm for dense and sparse linear assignment problems." Computing 38(4): 325-340.

Juang, C. F., & Chang, P. H. (2010). Designing Fuzzy-Rule-Based Systems Using Continuous Ant-Colony Optimization. Ieee Transactions on Fuzzy Systems, 18(1), 138-149. doi: 10.1109/tfuzz.2009.2038150

Kaveh, A., & Talatahari, S. (2010). An improved ant colony optimization for constrained engineering design problems. Engineering Computations, 27(1-2), 155-182. doi: 10.1108/02644401011008577

Kaveh, A., & Talatahari, S. (2010). An improved ant colony optimization for the design of planar steel frames. Engineering Structures, 32(3), 864-873. doi: 10.1016/j.engstruct.2009.12.012

Kaviri, A. G., Jaafar, M. N. M., & Lazim, T. M. (2012). Modeling and multi-objective exergy based optimization of a combined cycle power plant using a genetic algorithm. Energy Conversion and Management, 58, 94-103. doi: 10.1016/j.enconman.2012.01.002

Koshy, Thomas. Discrete Mathematics with Applications. Burlington, MA, USA: Academic Press 2003. p 746.

Kosmas, O. T., & Vlachos, D. S. (2012). Simulated annealing for optimal ship routing. Computers & Operations Research, 39(3), 576-581. doi: 10.1016/j.cor.2011.05.010

Kuo, R. J., & Han, Y. S. (2011). A hybrid of genetic algorithm and particle swarm optimization for solving bi-level linear programming problem - A case study on supply chain model. Applied Mathematical Modelling, 35(8), 3905-3917. doi: 10.1016/j.apm.2011.02.008

Kuo, Y. Y. (2010). Using simulated annealing to minimize fuel consumption for the time-dependent vehicle routing problem. Computers & Industrial Engineering, 59(1), 157-165. doi: 10.1016/j.cie.2010.03.012

Lee, C. Y., Lee, Z. J., Lin, S. W., & Ying, K. C. (2010). An enhanced ant colony optimization (EACO) applied to capacitated vehicle routing problem. Applied Intelligence, 32(1), 88-95. doi: 10.1007/s10489-008-0136-9

Lee, W. C., Chen, S. K., & Wu, C. C. (2010). Branch-and-bound and simulated annealing algorithms for a two-agent scheduling problem. Expert Systems with Applications, 37(9), 6594-6601. doi: 10.1016/j.eswa.2010.02.125

Leung, C. W., Wong, T. N., Mak, K. L., & Fung, R. Y. K. (2010). Integrated process planning and scheduling by an agent-based ant colony optimization. Computers & Industrial Engineering, 59(1), 166-180. doi: 10.1016/j.cie.2009.09.003

Li, X. D. (2010). Niching Without Niching Parameters: Particle Swarm Optimization Using a Ring Topology. Ieee Transactions on Evolutionary Computation, 14(1), 150-169. doi: 10.1109/tevc.2009.2026270

Lin, Y. T., Huang, Y. M., & Cheng, S. C. (2010). An automatic group composition system for composing collaborative learning groups using enhanced particle swarm optimization. Computers & Education, 55(4), 1483-1493. doi: 10.1016/j.compedu.2010.06.014

Liu, H., Cai, Z. X., & Wang, Y. (2010). Hybridizing particle swarm optimization with differential evolution for constrained numerical and engineering optimization. Applied Soft Computing, 10(2), 629-640. doi: 10.1016/j.asoc.2009.08.031

Magnier, L., & Haghighat, F. (2010). Multiobjective optimization of building design using TRNSYS simulations, genetic algorithm, and Artificial Neural Network. Building and Environment, 45(3), 739-746. doi: 10.1016/j.buildenv.2009.08.016

Meng, K., Wang, H. G., Dong, Z. Y., & Wong, K. P. (2010). Quantum-Inspired Particle Swarm Optimization for Valve-Point Economic Load Dispatch. Ieee Transactions on Power Systems, 25(1), 215-222. doi: 10.1109/tpwrs.2009.2030359

Mohan, B. C., & Baskaran, R. (2012). A survey: Ant Colony Optimization based recent research and implementation on several engineering domain. Expert Systems with Applications, 39(4), 4618-4627. doi: 10.1016/j.eswa.2011.09.076

Moradi, M. H., & Abedini, M. (2012). A combination of genetic algorithm and particle swarm optimization for optimal DG location and sizing in distribution systems. International Journal of Electrical Power & Energy Systems, 34(1), 66-74. doi: 10.1016/j.ijepes.2011.08.023

Moradi, M. H., & Abedini, M. (2012). A combination of genetic algorithm and particle swarm optimization for optimal DG location and sizing in distribution systems. International Journal of

Electrical Power & Energy Systems, 34(1), 66-74. doi: 10.1016/j.ijepes.2011.08.023

Moslehi, G., & Mahnam, M. (2011). A Pareto approach to multi-objective flexible job-shop scheduling problem using particle swarm optimization and local search. International Journal of Production Economics, 129(1), 14-22. doi: 10.1016/j.ijpe.2010.08.004

Musa, R., Arnaout, J. P., & Jung, H. (2010). Ant colony optimization algorithm to solve for the transportation problem of cross-docking network. Computers & Industrial Engineering, 59(1), 85-92. doi: 10.1016/j.cie.2010.03.002

Naderi, B., Tavakkoli-Moghaddam, R., & Khalili, M. (2010). Electromagnetism-like mechanism and simulated annealing algorithms for flowshop scheduling problems minimizing the total weighted tardiness and makespan. Knowledge-Based Systems, 23(2), 77-85. doi: 10.1016/j.knosys.2009.06.002

Nazghelichi, T., Aghbashlo, M., & Kianmehr, M. H. (2011). Optimization of an artificial neural network topology using coupled response surface methodology and genetic algorithm for fluidized bed drying. Computers and Electronics in Agriculture, 75(1), 84-91. doi: 10.1016/j.compag.2010.09.014

Neumann, F., & Witt, C. (2010). Ant Colony Optimization and the minimum spanning tree problem. Theoretical Computer Science, 411(25), 2406-2413. doi: 10.1016/j.tcs.2010.02.012

Nickabadi, A., Ebadzadeh, M. M., & Safabakhsh, R. (2011). A novel particle swarm optimization algorithm with adaptive inertia weight. Applied Soft Computing, 11(4), 3658-3670. doi: 10.1016/j.asoc.2011.01.037

Niknam, T. (2010). A new fuzzy adaptive hybrid particle swarm optimization algorithm for non-linear, non-smooth and non-convex economic dispatch problem. Applied Energy, 87(1), 327-339. doi: 10.1016/j.apenergy.2009.05.016

Niknam, T., Mojarrad, H. D., & Nayeripour, M. (2010). A new fuzzy adaptive particle swarm optimization for non-smooth economic dispatch. Energy, 35(4), 1764-1778. doi: 10.1016/j.energy.2009.12.029

Niu, D. X., Wang, Y. L., & Wu, D. D. (2010). Power load forecasting using support vector machine and ant colony optimization. Expert Systems with Applications, 37(3), 2531-2539. doi: 10.1016/j.eswa.2009.08.019

Ozcan, U. (2010). Balancing stochastic two-sided assembly lines: A chance-constrained, piecewise-linear, mixed integer program and a simulated annealing algorithm. European Journal of Operational Research 205(1), 81-97. doi: 10.1016/j.ejor.2009.11.033

Park, J. B., Jeong, Y. W., Shin, J. R., & Lee, K. Y. (2010). An Improved Particle Swarm Optimization for Nonconvex Economic Dispatch Problems. Ieee Transactions on Power Systems, 25(1), 156-166. doi: 10.1109/tpwrs.2009.2030293

Pedemonte, M., Nesmachnow, S., & Cancela, H. (2011). A survey on parallel ant colony optimization. Applied Soft Computing, 11(8), 5181-5197. doi: 10.1016/j.asoc.2011.05.042

Pedersen, M. E. H., & Chipperfield, A. J. (2010). Simplifying Particle Swarm Optimization. Applied Soft Computing, 10(2), 618-628. doi: 10.1016/j.asoc.2009.08.029

Pishvaee, M. S., Kianfar, K., & Karimi, B. (2010). Reverse logistics network design using simulated annealing. International Journal of Advanced Manufacturing Technology, 47(1-4), 269-281. doi: 10.1007/s00170-009-2194-5

Poli, L., Rocca, P., Manica, L., & Massa, A. (2010). Handling Sideband Radiations in Time-Modulated Arrays Through Particle Swarm Optimization. Ieee Transactions on Antennas and Propagation, 58(4), 1408-1411. doi: 10.1109/tap.2010.2041165

Precup, R. E., David, R. C., Petriu, E. M., Preitl, S., & Radac, M. B. (2012). Fuzzy Control Systems With Reduced Parametric Sensitivity Based on Simulated Annealing. Ieee Transactions on Industrial Electronics, 59(8), 3049-3061. doi: 10.1109/tie.2011.2130493

Putha, R., Quadrifoglio, L., & Zechman, E. (2012). Comparing Ant Colony Optimization and Genetic Algorithm Approaches for Solving Traffic Signal Coordination under Oversaturation Conditions. Computer-Aided Civil and Infrastructure Engineering, 27(1), 14-28. doi: 10.1111/j.1467-8667.2010.00715.x

Putha, R., Quadrifoglio, L., & Zechman, E. (2012). Comparing Ant Colony Optimization and Genetic Algorithm Approaches for Solving Traffic Signal Coordination under Oversaturation Conditions. Computer-Aided Civil and Infrastructure Engineering, 27(1), 14-28. doi: 10.1111/j.1467-8667.2010.00715.x

Qin, X. S., Huang, G. H., & Liu, L. (2010). A Genetic-Algorithm-Aided Stochastic Optimization Model for Regional Air Quality Management under Uncertainty. Journal of the Air & Waste Management Association, 60(1), 63-71. doi: 10.3155/1047-

3289.60.1.63

Queirolo, C. C., Silva, L., Bellon, O. R. P., & Segundo, M. P. (2010). 3D Face Recognition Using Simulated Annealing and the Surface Interpenetration Measure. Ieee Transactions on Pattern Analysis and Machine Intelligence, 32(2) 206-219. doi: 10.1109/tpami.2009.14

R. Venkata Rao; Vimal J. Savsani. (2012). Mechanical Design Optimization Using Advanced Optimization Techniques. Springer London

Rubinstein, Reuven Y.; Ridder, Ad; Vaisman, Radislav. Wiley Series in Probability and Statistics: Fast Sequential Monte Carlo Methods for Counting and Optimization. Somerset, NJ, USA: Wiley 2013. p 6.

Sanaye, S., & Hajabdollahi, H. (2010). Thermal-economic multi-objective optimization of plate fin heat exchanger using genetic algorithm. Applied Energy, 87(6), 1893-1902. doi: 10.1016/j.apenergy.2009.11.016

Sandou, Guillaume. FOCUS Series : Metaheuristic Optimization for the Design of Automatic Control Laws. Somerset, NJ, USA: Wiley 2013. p 13.

Sandou, Guillaume. FOCUS Series : Metaheuristic Optimization for the Design of Automatic Control Laws. Somerset, NJ, USA: Wiley 2013. p 9.

Santos, L., Coutinho-Rodrigues, J., & Current, J. R. (2010). An improved ant colony optimization based algorithm for the capacitated arc routing problem. Transportation Research Part B-Methodological, 44(2), 246-266. doi: 10.1016/j.trb.2009.07.004

Shankar, T. J., Sokhansanj, S., Bandyopadhyay, S., & Bawa, A. S. (2010). A Case Study on Optimization of Biomass Flow During Single-Screw Extrusion Cooking Using Genetic Algorithm (GA) and Response Surface Method (RSM). Food and Bioprocess Technology, 3(4), 498-510. doi: 10.1007/s11947-008-0172-9

Sharma, A. K., Son, K. H., Han, B. Y., & Sohn, K. S. (2010). Simultaneous Optimization of Luminance and Color Chromaticity of Phosphors Using a Nondominated Sorting Genetic Algorithm. Advanced Functional Materials 20(11), 1750-1755. doi: 10.1002/adfm.200902285

Shi, Y., Liu, H. C., Gao, L., & Zhang, G. H. (2011). Cellular particle swarm optimization. Information Sciences, 181(20), 4460-4493. doi: 10.1016/j.ins.2010.05.025

Singh, H. K., Ray, T., & Smith, W. (2010). C-PSA: Constrained Pareto simulated annealing for constrained multi-objective optimization.

Information Sciences, 180(13), 2499-2513. doi: 10.1016/j.ins.2010.03.021

Sivapathasekaran, C., Mukherjee, S., Ray, A., Gupta, A., & Sen, R. (2010). Artificial neural network modeling and genetic algorithm based medium optimization for the improved production of marine biosurfactant. Bioresource Technology, 101(8), 2884-2887. doi: 10.1016/j.biortech.2009.09.093

Sousa, T., Morais, H., Vale, Z., Faria, P., & Soares, J. (2012). Intelligent Energy Resource Management Considering Vehicle-to-Grid: A Simulated Annealing Approach. Ieee Transactions on Smart Grid, 3(1), 535-542. doi: 10.1109/tsg.2011.2165303

Stivala, A. D., Stuckey, P. J., & Wirth, A. I. (2010). Fast and accurate protein substructure searching with simulated annealing and GPUs. Bmc Bioinformatics, 11. doi: 10.1186/1471-2105-11-446

Tatipamula, Mallikarjun; Oki, Eiji; Rojas-Cessa, Roberto. Advanced Internet Protocols, Services, and Applications. Hoboken, NJ, USA: Wiley 2012. p 48.

Tian, J., Ma, L. H., & Yu, W. Y. (2011). Ant colony optimization for wavelet-based image interpolation using a three-component exponential mixture model. Expert Systems with Applications, 38(10), 12514-12520. doi: 10.1016/j.eswa.2011.04.037

Tycko, R., & Hu, K. N. (2010). A Monte Carlo/simulated annealing algorithm for sequential resonance assignment in solid state NMR of uniformly labeled proteins with magic-angle spinning. Journal of Magnetic Resonance 205(2), 304-314. doi: 10.1016/j.jmr.2010.05.013

Unler, A., & Murat, A. (2010). A discrete particle swarm optimization method for feature selection in binary classification problems. European Journal of Operational Research 206(3), 528-539. doi: 10.1016/j.ejor.2010.02.032

Valdez, F., Melin, P., & Castillo, O. (2011). An improved evolutionary method with fuzzy logic for combining Particle Swarm Optimization and Genetic Algorithms. Applied Soft Computing, 11(2), 2625-2632. doi: 10.1016/j.asoc.2010.10.010

Wai, R. J., Lee, J. D., & Chuang, K. L. (2011). Real-Time PID Control Strategy for Maglev Transportation System via Particle Swarm Optimization. Ieee Transactions on Industrial Electronics, 58(2), 629-646. doi: 10.1109/tie.2010.2046004

Wang, J. F., Sun, Z. X., Dai, Y. P., & Ma, S. L. (2010). Parametric optimization design for supercritical CO_2 power cycle using genetic algorithm and artificial neural network. Applied Energy, 87(4),

1317-1324. doi: 10.1016/j.apenergy.2009.07.017

Wang, J. J., Jing, Y. Y., & Zhang, C. F. (2010). Optimization of capacity and operation for CCHP system by genetic algorithm. Applied Energy, 87(4), 1325-1335. doi: 10.1016/j.apenergy.2009.08.005

Wang, J. J., Zhai, Z. Q., Jing, Y. Y., & Zhang, C. F. (2010). Particle swarm optimization for redundant building cooling heating and power system. Applied Energy, 87(12), 3668-3679. doi: 10.1016/j.apenergy.2010.06.021

Wang, Y. C., Lv, J. A., Zhu, L., & Ma, Y. M. (2010). Crystal structure prediction via particle-swarm optimization. Physical Review B, 82(9). doi: 10.1103/PhysRevB.82.094116

Wang, Y., Li, B., Weise, T., Wang, J. Y., Yuan, B., & Tian, Q. J. (2011). Self-adaptive learning based particle swarm optimization. Information Sciences, 181(20), 4515-4538. doi: 10.1016/j.ins.2010.07.013

Wee, Hui-Ming. Management Science - Theory and Applications : Inventory Systems : Modeling and Research Methods. New York, NY, USA: Nova 2011. p 83.

Woodruff, David L. (Editor). Network Interdiction and Stochastic Integer Programming. Secaucus, NJ, USA: Kluwer Academic Publishers 2002. p 99.

Wu, D. F., Warwick, K., Ma, Z., Gasson, M. N., Burgess, J. G., Pan, S., & Aziz, T. Z. (2010). Prediction Of Parkinson's Disease Tremor Onset Using A Radial Basis Function Neural Network Based On Particle Swarm Optimization. International Journal of Neural Systems 20(2), 109-116. doi: 10.1142/s0129065710002292

Wu, Q. (2010). A hybrid-forecasting model based on Gaussian support vector machine and chaotic particle swarm optimization. Expert Systems with Applications, 37(3), 2388-2394. doi: 10.1016/j.eswa.2009.07.057

Wu, W. Y., Simpson, A. R., & Maier, H. R. (2010). Accounting for Greenhouse Gas Emissions in Multiobjective Genetic Algorithm Optimization of Water Distribution Systems. Journal of Water Resources Planning and Management-Asce, 136(2), 146-155. doi: 10.1061/(asce)wr.1943-5452.0000020

Xavier-de-Souza, S., Suykens, J. A. K., Vandewalle, J., & Bolle, D. (2010). Coupled Simulated Annealing. Ieee Transactions on Systems Man and Cybernetics Part B-Cybernetics, 40(2), 320-335. doi: 10.1109/tsmcb.2009.2020435

Xing, L. N., Chen, Y. W., Wang, P., Zhao, Q. S., & Xiong, J. (2010). Knowledge-Based Ant Colony Optimization for Flexible Job Shop Scheduling Problems. Applied Soft Computing, 10(3), 888-896. doi: 10.1016/j.asoc.2009.10.006

Xu, T., Zuo, W. J., Xu, T. S., Song, G. C., & Li, R. C. (2010). An adaptive reanalysis method for genetic algorithm with application to fast truss optimization. Acta Mechanica Sinica, 26(2), 225-234. doi: 10.1007/s10409-009-0323-x

Yang, J. G., & Zhuang, Y. B. (2010). An improved ant colony optimization algorithm for solving a complex combinatorial optimization problem. Applied Soft Computing, 10(2), 653-660. doi: 10.1016/j.asoc.2009.08.040

Yang, Xin-She. Engineering Optimization: An Introduction with Metaheuristic Applications. Hoboken, NJ, USA: Wiley 2010. p 173-190.

Yang, Xin-She. Introduction to Mathematical Optimization: From Linear Programming to Metaheuristics. Cambridge, GBR: Cambridge International Science Publishing 2008. p 100-119.

Yu, B., & Yang, Z. Z. (2011). An ant colony optimization model: The period vehicle routing problem with time windows. Transportation Research Part E-Logistics and Transportation Review, 47(2), 166-181. doi: 10.1016/j.tre.2010.09.010

Yu, B., Yang, Z. Z., & Xie, J. X. (2011). A parallel improved ant colony optimization for multi-depot vehicle routing problem. Journal of the Operational Research Society, 62(1), 183-188. doi: 10.1057/jors.2009.161

Yu, V. F., Lin, S. W., Lee, W., & Ting, C. J. (2010). A simulated annealing heuristic for the capacitated location routing problem. Computers & Industrial Engineering, 58(2), 288-299. doi: 10.1016/j.cie.2009.10.007

Zhan, Z. H., Zhang, J., Li, Y., & Shi, Y. H. (2011). Orthogonal Learning Particle Swarm Optimization. Ieee Transactions on Evolutionary Computation, 15(6), 832-847. doi: 10.1109/tevc.2010.2052054

Zhang, B., Sun, X., Gao, L. R., & Yang, L. N. (2011). Endmember Extraction of Hyperspectral Remote Sensing Images Based on the Ant Colony Optimization (ACO) Algorithm. Ieee Transactions on Geoscience and Remote Sensing, 49(7), 2635-2646. doi: 10.1109/tgrs.2011.2108305

Zhang, R., & Wu, C. (2010). A hybrid immune simulated annealing algorithm for the job shop scheduling problem. Applied Soft Computing, 10(1), 79-89. doi: 10.1016/j.asoc.2009.06.008

Zhang, R., & Wu, C. (2011). A simulated annealing algorithm based on block properties for the job shop scheduling problem with total weighted tardiness objective. Computers & Operations Research, 38(5), 854-867. doi: 10.1016/j.cor.2010.09.014

Zhang, Y., Xu, J. L., Yuan, Z. H., Xu, H. J., & Yu, Q. (2010). Artificial neural network-genetic algorithm based optimization for the immobilization of cellulase on the smart polymer Eudragit L-100. Bioresource Technology, 101(9), 3153-3158. doi: 10.1016/j.biortech.2009.12.080

Part II

Chapter 4. Reducing the Kullback-Leibler Distance: The Cross-Entropy Method for Optimization

Abstract. This study presents a brief applied optimization technique, which utilizes the cross entropy (CE) method for finding minimum/maximum values of selected optimization problems. By utilizing the CE method, optimum solutions are achieved for sample optimization problems.

Keywords: Kullback-Leibler distance, cross-entropy method, optimization

1. Introduction

Rubinstein developed the Cross-Entropy (CE) method in 1997 and it is adapted for combinatorial optimization solutions (Rubinstein 1997, 1999, 2001; Rubinstein and Kroese 2004; Rubinstein and Melamed, 1998; Rubinstein and Shapiro 1993). The idea behind the CE method is to model an effective learning technique throughout the search process of the algorithm to solve combinatorial optimization problems. The method first produces a random sample from a pre-specified probability distribution function and then treats the sample to adjust the parameters of the probability distribution in order to generate a better sample in the next iteration.

The remainder of this study is organized as follows. Section 2 briefly reviews the literature on CE method. Section 3 introduces the method used for optimization. Section 4 presents some example functions for finding global minimum. The study is concluded in Section 5.

2. Literature review

The significance of the CE method is that it defines a precise mathematical framework for deriving fast, and in some sense "optimal" updating/learning rules, based on advanced simulation theory (de Boer *et al.* 2005).

While most of the stochastic algorithms for combinatorial optimization are based on local search (they employ local neighborhood structures), the cross-entropy method is a global random search procedure (Margolin 2005). The method consists of an iterative stochastic procedure that makes use of the importance sampling technique (Margolin 2005).

There are many examples for solving optimization problems using CE method. For instance, the literature about solving problems for vehicle routing (Chepuri and Homem-de-Mello, 2005), for max-cut and bipartition problems (Rubinstein, 2002), for project management (Cohen, Golany, and Shtub, 2005) and for scheduling (Margolin, 2002, 2004). Some studies that use cross entropy method are highlighted in Table 1 through 5.

Table 1. The CE method literature (Mathematics and statistics)

An improved cross-entropy method applied to inverse problems (an *et al.* 2012).
The generalized cross entropy method, with applications to probability density estimation (Botev and Kroese 2011).
Improved cross-entropy method for estimation (Chan and Kroese 2012).
The cross-entropy method and its application to inverse problems (Ho and Yang 2010).
Multiobjective optimization of inverse problems using a vector cross entropy method (Ho and Yang 2012).

Table 2. The CE method literature (Computer science)

Design of a new interleaver using cross entropy method for turbo coding (Abderrahmane 2013).
A self-organization mechanism based on cross-entropy method for p2p-like applications (Chen *et al.* 2010).
Multi-objective buffer space allocation with the cross-entropy method (Bekker 2013).

Table 3. The CE method literature (Electronics and manufacturing)

Efficient capacity-based joint quantized precoding and transmit antenna selection using cross-entropy method for multiuser MIMO systems (Xhen *et al.* 2012).
Peak power reduction of OFDM systems through tone injection via parametric minimum cross-entropy method (Damavandi *et al.* 2013).
On the benefits of laplace samples in solving a rare event problem using cross-entropy method (Selvan *et al.* 2013).
Shaped beam synthesis of phased arrays using the cross entropy method (Weatherspoon *et al.* 2013).
Cross-entropy method for the optimization of optical alignment signals with diffractive effects (Chen *et al.* 2011).
Signal optimisation using the cross entropy method (Maher *et al.* 2013).
Sparse antenna array optimization with the cross-entropy method (Minvielle *et al.* 2011).
The cross-entropy method and its application to minimize the ripple of magnetic levitation forces of a maglev system (Zhang *et al.* 2010).

Table 4. The CE method literature (Optimization problems)

The cross-entropy method for combinatorial optimization problems of seaport logistics terminal (Yildiz and Yercan 2010).
Solving the multidimensional assignment problem by a cross-entropy method (Nguyen *et al.* 2014).
Simulation optimization using the cross-entropy method with optimal computing budget allocation (He *et al.* 2010).
The cross-entropy method in multi-objective optimisation: an assessment (Bekker and Aldrich 2011).

Table 5. The CE method literature (Others)

A suboptimal tone reservation algorithm based on cross-entropy method for PAPR reduction in OFDM systems (Chen *et al.* 2011).
Design of two-dimensional zero reference codes with cross-entropy method (Chen and Wen 2010).
Estimating change-points in biological sequences via the cross-entropy method (Evans *et al.* 2011).
A combined splitting-cross entropy method for rare-event probability estimation of queueing networks (Garvels 2011).
Optimal fuzzy control system using the cross-entropy method. A case study of a drilling process (Haber *et al.* 2010).
Cooperative cross-entropy method for generating entangled networks (Hui 2011).
The cross-entropy method with patching for rare-event simulation of large markov chains (Kaynar and Ridder 2010).
User selection method adopting cross-entropy method for a downlink multiuser MIMO system (Kim at al 2014).
The cross-entropy method for reliability assessment of cracked structures subjected to random markovian loads (Mattrand and Bourinet 2014).
Calibration of second order traffic models using continuous cross entropy method (Ngoduy and Maher 2012).
Stochastic inversion of ocean color data using the cross-entropy method (Salama and Shen 2010).
Enhanced cross-entropy method for dynamic economic dispatch with valve-point effects (Selvakumar 2011).
Fast reconstruction of computerized tomography images based on the cross-entropy method (Wang *et al.* 2011).

3. The CE Method

The global maximum of function $S(x)$ is represented by,

$$S(x^*) = \gamma^* = \max_{x \in X}(S(x))$$

The probability that the score function $S(x)$ evaluated at a particular state x is close to γ^* is classified as rare-event.

$$P_v(S(x) \geq \gamma) = E_v I_{\{S(x) \geq \gamma\}}$$

The important element of *maximum likelihood estimation* (MLE) is that there is a definable probability function that could be used to generate the likelihood of the observed event. It is shown as,

$$\hat{v}^* = \arg\max_v \left((1/N_s) \sum_{i=1}^{N_s} I_{\{S(x) \geq \gamma\}} \ln \varphi(x_i, v) \right)$$

The above statement comes from the definition of Kullback-Leibler (KL) distance. The Kullback-Leibler distance or cross entropy is a measure of the distance between two probability distributions $D_{KL}(P,Q)$. The stochastic optimization problem is solved by identifying the optimal *importance sampling* (IS) density that minimizes KL distance regarding the original density function. KL distance is the cross entropy between the original density function and the importance sampling density function. The distance $D_{KL}(g,h)$ is determined as a particular suitable criterion between densities of g and h. The KL distance (cross-entropy) is

$$D_{KL}(P,Q) = E_P \ln \frac{P(x)}{Q(x)} = \int P(x) \ln P(x) dx - \int P(x) \ln Q(x) dx$$

For discrete variables, the Kullback-Leibler (KL) divergence of Q from P is,

$$D_{KL}(P,Q) = \sum_x P(x) \log(\frac{P(x)}{Q(x)}) = \sum_x P(x) \log(P(x)) - \sum_x P(x) \log(Q(x))$$

For continuous variables, the Kullback-Leibler (KL) divergence of Q from P is,

$$D_{KL}(P,Q) = \int_{-\infty}^{+\infty} P(x) \ln(\frac{P(x)}{Q(x)}) dx = \int_{-\infty}^{+\infty} P(x) \ln(P(x)) dx - \int_{-\infty}^{+\infty} P(x) \ln(Q(x)) dx$$

$$\text{where } D_{KL}(P,Q) = H(P,Q) - H(P)$$

$H(P,Q)$ is named as the cross-entropy between P and Q. $H(P)$ is the entropy of P. The minimization of the KL distance (cross-entropy) provides definition for the parameters of the density functions and generations of enhanced feasible vectors. The method aborts when it comes together into a solution in the feasible region.

$$D_{KL}(P,Q) = E_P \ln \frac{P(x)}{Q(x)}$$

$$D_{KL}(P,Q) = \int P(x) \ln P(x) dx - \int P(x) \ln Q(x) dx$$

Minimizing the KL divergence between $P(x)$ and $Q(x)$ depends on the second term of the above statement, where $Q(x) = \varphi(x,V)$, then the equivalent maximization statement is,

$$\min_V \left(-\int P^*(x) \ln \varphi(x,V) dx \right) = \max_V \left(\int P^*(x) \ln \varphi(x,V) dx \right)$$

$$P^*(x) = \frac{I_{\{S(x) \geq \gamma\}} \varphi(x,u)}{l}$$

By substituting $P^*(x)$ in to the statement,

$$\max_V \left(\int \frac{I_{\{S(x) \geq \gamma\}} \varphi(x,u)}{l} \ln \varphi(x,V) dx \right)$$

The equivalent statement with an expectation E operator is,

$$\max_V D(V) = \max_V \left(E_u I_{\{S(x) \geq \gamma\}} \ln \varphi(x,V) dx \right)$$

3.1 The continuous CE optimization

The normal distribution $N(\mu, \sigma^2)$ function $\varphi(x)$ is,

$$\varphi(x) = \frac{1}{\sqrt{2\pi\sigma^2}} e^{-\frac{1}{2}\left(\frac{x-\mu}{\sigma}\right)^2}, \quad x \in R$$

By taking natural logarithms of both sides of the distribution function,

$$\ln(\varphi(x)) = \ln\left(\frac{1}{\sqrt{2\pi\sigma^2}} e^{-\frac{1}{2}\left(\frac{x-\mu}{\sigma}\right)^2}\right)$$

Thus, the statement becomes,

$$\ln(\varphi(x)) = -\left(\frac{1}{2}\ln(2\pi) + \frac{1}{2}\ln(\sigma^2) + \frac{1}{2}\left(\frac{x-\mu}{\sigma}\right)^2\right)$$

the maximum log-likelihood estimator is,

$$\hat{v}^* = \arg\max_{v} \left((1/N_s) \sum_{i=1}^{N_s} I_{\{S(x) \geq \gamma\}} \ln \varphi(x_i, v)\right)$$

By substituting the function $\varphi(x)$ into the maximum log-likelihood estimator,

$$\hat{v}^* = \arg\max_{v} \left((1/N_s) \sum_{i=1}^{N_s} I_{\{S(x) \geq \gamma\}} \left[-\left(\frac{1}{2}\ln(2\pi) + \frac{1}{2}\ln(\sigma^2) + \frac{1}{2}\left(\frac{x_i-\mu}{\sigma}\right)^2\right)\right]\right)$$

The maximization problem becomes a minimization, since from inspection of the above statement,

$$-\max(-f(x)) = \min(f(x))$$

Then,

$$\hat{\tau}^* = \min_\tau \left((1/2N_s) \sum_{i=1}^{N_s} I_{\{S(x) \geq \gamma\}} \left[\left(\ln(2\pi) + \ln(\sigma^2) + \left(\frac{x-\mu}{\sigma}\right)^2 \right) \right] \right)$$

By eliminating the non-variable parts, the minimization can be simplified as,

$$\min \left(\ln(\sigma^2) \sum_{i=1}^{N_s} I_i + \left(\frac{1}{\sigma^2}\right) \sum_{i=1}^{N_s} I_i (x_i - \mu)^2 \right)$$

By taking partial derivatives with respect to μ,

$$\frac{\partial}{\partial \mu} \left(\ln(\sigma^2) \sum_{i=1}^{N_s} I_i + \left(\frac{1}{\sigma^2}\right) \sum_{i=1}^{N_s} I_i (x_i - \mu)^2 \right) =$$

$$\frac{2}{\sigma^2} \left(\sum_{i=1}^{N_s} I_i \mu - \sum_{i=1}^{N_s} I_i x_i \right) = 0$$

$$\frac{2}{\sigma^2} \sum_{i=1}^{N_s} I_i \mu = \frac{2}{\sigma^2} \sum_{i=1}^{N_s} I_i x$$

Moreover, placing μ on the left hand side, the following estimate is obtained,

$$\hat{\mu} = \frac{\sum_{i=1}^{N_s} I_i x}{\sum_{i=1}^{N_s} I_i}$$

By taking partial derivatives with respect to σ^2,

$$\frac{\partial}{\partial \sigma^2}\left(\ln(\sigma^2)\sum_{i=1}^{N_s} I_i + \left(\frac{1}{\sigma^2}\right)\sum_{i=1}^{N_s} I_i(x_i - \mu)^2 \right) =$$

$$\frac{1}{\sigma^2}\sum_{i=1}^{N_s} I_i - \frac{1}{(\sigma^2)^2}\sum_{i=1}^{N_s} I_i(x_i - \mu)^2 = 0$$

$$\sum_{i=1}^{N_s} I_i = \frac{1}{(\sigma^2)}\sum_{i=1}^{N_s} I_i(x_i - \mu)^2$$

By placing σ^2 on the left hand side, the following estimate is obtained,

$$\hat{\sigma}^2 = \frac{\sum_{i=1}^{N_s} I_i(x_i - \mu)^2}{\sum_{i=1}^{N_s} I_i}$$

3.2 The combinatorial CE optimization

The random vector is $X = (X_1, ..., X_n) \sim Ber(p)$. Density function φ on X parameterized by a vector $p \in [0,1]^n$. Bernoulli density function, under the following probability density function (pdf) is,

$$\varphi(x, p) = \prod_{i=1}^{n}(p_i)^{x_i}(1 - p_i)^{1-x_i} \quad ,$$

$$x \in \{0,1\}$$

which is,

$$\varphi(x; p) = \begin{cases} p & \text{if } x = 1, \\ 1 - p & \text{if } x = 0, \\ 0 & \text{otherwise} \end{cases}$$

$$P_v(S(x) \geq \gamma) = E_v I_{\{S(x) \geq \gamma\}}$$

By taking natural logarithms of both sides of the distribution function $\varphi(x; p)$,

$$\ln \varphi(x; p) = \ln\left(p^x (1-p)^{1-x}\right)$$

yields,

$$\ln \varphi(x; p) = x \ln p + (1-x)\ln(1-p)$$

By taking partial derivatives with respect to p,

$$\frac{\partial}{\partial p} \ln \varphi(x; p) = \frac{x_i}{p} - \frac{1-x_i}{1-p}$$

The statement becomes,

$$\frac{\partial}{\partial p} \ln \varphi(x; p) = \frac{x_i - p}{p(1-p)}$$

The maximum log-likelihood is,

$$\hat{v}^* = \arg\max_{v} \left((1/N_s) \sum_{i=1}^{N_s} I_{\{S(x) \geq \gamma\}} \ln \varphi(x_i, v) \right)$$

By substituting,

$$\frac{\partial}{\partial p}\left(\sum_{i=1}^{N_s} I_{\{S(x) \geq \gamma\}} \ln \varphi(x_i, v) \right) = \sum_{i=1}^{N_s} I_{\{S(x) \geq \gamma\}} \frac{x_i - p}{p(1-p)}$$

$$= (1/p(1-p)) \sum_{i=1}^{N_s} I_{\{S(x) \geq \gamma\}} (x_i - p)$$

In addition, placing p on the left hand side, the following estimate is obtained,

$$\hat{p} = \frac{\sum_{i=1}^{N_s} I_{\{S(x) \geq \gamma\}} x_i}{\sum_{i=1}^{N_s} I_{\{S(x) \geq \gamma\}}}$$

The CE method involves an iterative procedure where each iteration can be broken down into two phases (de Boer *et al.* 2005):

1. Generate a random data sample (trajectories, vectors, etc.) according to a specified mechanism.

2. Update the parameters of the random mechanism based on the data to produce a "better" sample in the next iteration.

The general cross-entropy method consists of three main steps (Margolin 2005):

1. Choosing a probability family. Initializing of the distribution parameters and the method parameters. Generating feasible solutions according to the chosen distribution.

2. Updating the distribution parameters, based on the Kullback-Leibler cross-entropy.

3. Checking the stopping rule. Updating the method parameters in the case the stopping rule fails.

A general view of the pseudo-code for the cross-entropy algorithm is (Connor 2008):

1. Initialize parameters: Set initial parameter $\hat{v}^{(0)}$, a small value of ρ, set population size K, smoothing constant α and set iteration counter t = 1

2. Update $\hat{\gamma}^{(t)}$: Given $\hat{v}^{(t-1)}$, let $\hat{\gamma}^{(t)}$ be the $(1-\rho)$- quantile of $Z(x)$ satisfying

$$P_{v^{(t-1)}}(Z(x) \geq \gamma^{(t)}) \geq \rho$$
$$P_{v^{(t-1)}}(Z(x) \leq \gamma^{(t)}) \geq 1-\rho$$

with x sampled from $f(\cdot, \hat{v}^{(t-1)})$. Then, the estimate of $\gamma^{(t)}$ is calculated computed as $\hat{\gamma}^{(t)} = Z_{(\lceil 1-\rho K \rceil)}$, where $\lceil \cdot \rceil$ rounds $(1-\rho)K$ towards infinity.

3. Update $\hat{v}^{(t)}$: Given $\hat{v}^{(t-1)}$, determine $\hat{v}^{(t)}$ by solving the CE program

$$\hat{v}^{(t)} = \max_{v} \frac{1}{N_s} \sum_{i=1}^{N_s} I_{\{Z(x_i) \geq \hat{\gamma}^{(t)}\}} \ln f(x_i, v)$$

4. Optional step: (Smooth update of $\hat{v}^{(t)}$) To decrease the probability of the CE procedure converging too quickly to a suboptimal solution, a smoothed update of $\hat{v}^{(t)}$ can be computed.

$$\hat{v}^{(t)} = \alpha \cup^{(t)} + (1-\alpha)\hat{v}^{(t-1)}$$

where $\cup^{(t)}$ is the estimate of the parameter vector computed with (3), $\hat{v}^{(t-1)}$ is the parameter estimate from the previous iteration and $\alpha (for\, 0 < \alpha \leq 1)$ is a constant smoothing coefficient. By setting $\alpha = 1$, the update will not be smoothed.

5. If convergence is reached then stop; otherwise, Set t=t+1 and reiterate from step 2 to 4, until the stopping criteria is satisfied.

4. Examples

In this section, some test functions are used to evaluate the characteristics of the CE algorithm's search for the global minimum. These test functions are the Giunta, Ackley, Rastrigin, and Hölder table. Each subsection demonstrates the CE method's search for the global minimum.

4.1 Finding the minimum value of the Giunta test function

In this subsection, a sample optimization is carried out on the Giunta test function (Giunta 1997) by utilizing the CE method. The Giunta test function is defined in the search domain $x_1, x_2 \in [-1,1]$ as,

$$f(x) = \frac{3}{5} + \sum_{i=1}^{2} [\sin(\frac{16}{15}x_i - 1) + \sin^2(\frac{16}{15}x_i - 1) + \frac{1}{50}\sin 4((\frac{16}{15}x_i - 1))]$$

The Giunta function is known that it has minimum for x_1, x_2 both having values of 0.45834282 (see Figure 1).

$$f_{\min}(x_1, x_2) = 0.0602472184$$

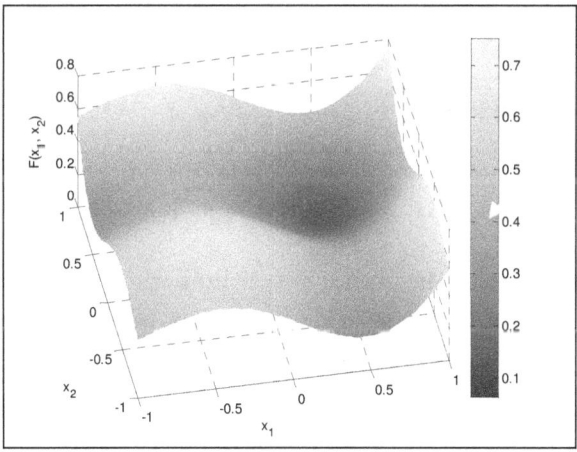

Figure 1. The plot of the Giunta function in 3D

The search for the optimum value is shown in the Figure 2. At each iteration, better vectors are created and each of these vectors are used to generate better values. Algorithm will stop when it converges to a global optimum value. The global minimum is shown with a cross symbol in the Figure 2.

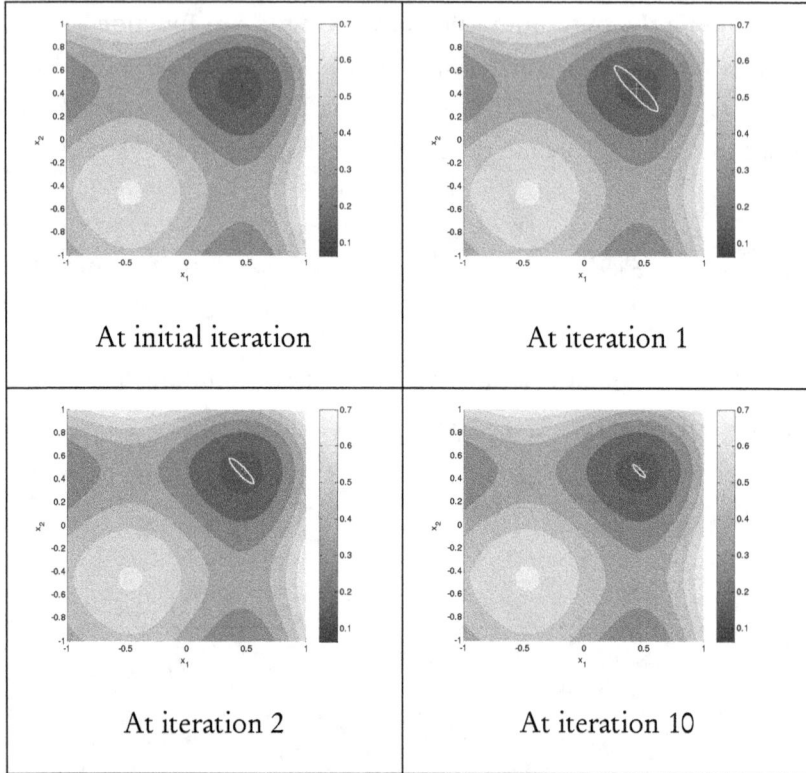

Figure 2. The contour plot of the Giunta test function – The search of a global minimum at each iteration.

In this example, with the author's own limitation to 10 iterations, the optimum value is found (see Table 6). At each iteration, algorithm tries better (x_1, x_2) values to obtain $f_{min}(x_1, x_2)$. At the 10th iteration, minimum value is

$$f_{min}(0.4673, 0.4673) = 0.0645$$

Table 6. CE algorithm's search for the optimum values

Iteration number	$\hat{\gamma}^{(t)}$	$\hat{v}^{(t)}$	x_1	x_2
1	0.0645	0.1043	0.3997	0.3956
2	0.0645	0.0666	0.4444	0.4441
3	0.0645	0.0645	0.4657	0.4653
4	0.0645	0.0645	0.4674	0.4674
5	0.0645	0.0645	0.4676	0.4673
6	0.0645	0.0645	0.4674	0.4673
7	0.0645	0.0645	0.4674	0.4673
8	0.0645	0.0645	0.4672	0.4674
9	0.0645	0.0645	0.4674	0.4673
10	0.0645	0.0645	0.4673	0.4673

4.2 Finding the minimum value of the Ackley test function

The Ackley test function f is defined as,

$$F(x1, x2) = 20 - 20\,e^{(-0.2\sqrt{0.5\,x1^2 + 0.5\,x2^2})} - e^{(0.5\cos(2\pi x1) + 0.5\cos(2\pi x2))} + e$$

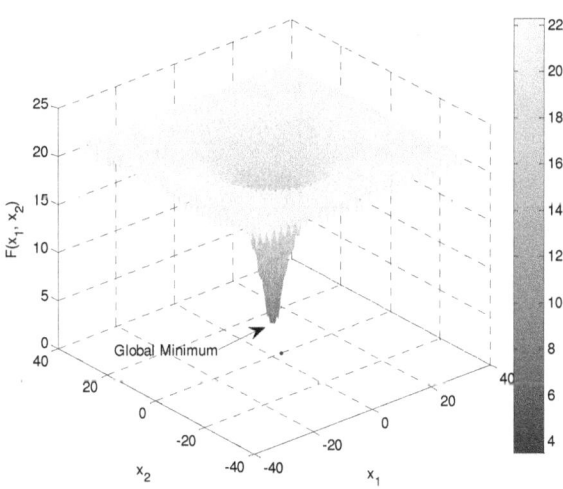

Figure 3. The plot of the Ackley function in 3D.

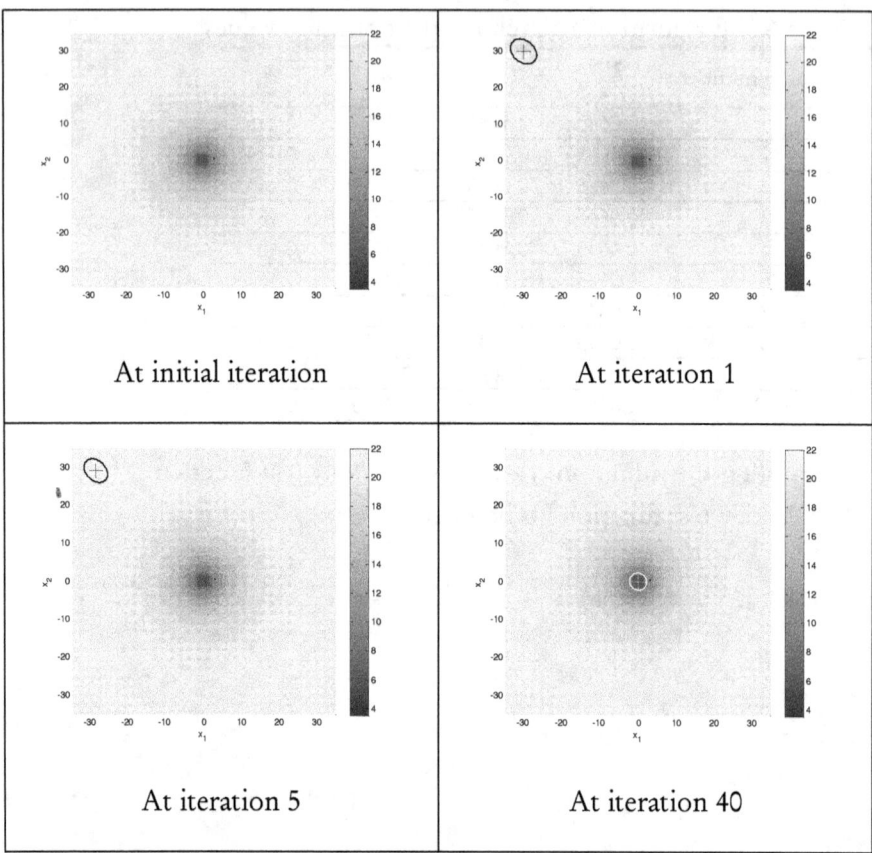

Figure 4. The contour plot of the Ackley test function – The search of a global minimum at each iteration.

Table 7. CE algorithm's search for the optimum values

Iteration number	$\hat{\gamma}^{(t)}$	$\hat{v}^{(t)}$	x_1	x_2
1	19.9222	19.9914	-29.7293	29.8061
2	19.9256	19.9884	-29.3937	29.5592
3	19.9029	19.9868	-29.1167	29.3436
4	19.8990	19.9820	-28.8572	29.1877
5	19.9092	19.9839	-28.5431	28.9334
6	19.9014	19.9810	-28.2489	28.6460
7	19.8937	19.9690	-27.9809	28.2785
8	19.8819	19.9730	-27.6325	27.8550
9	19.8829	19.9636	-27.2615	27.4680
10	19.8381	19.9542	-26.8110	27.1919
11	19.8603	19.9446	-26.2334	26.6838
12	19.8458	19.9400	-25.7706	26.2210
13	19.8364	19.9291	-25.2951	25.6356
14	19.8117	19.9131	-24.7197	24.9332
15	19.7725	19.8996	-24.1052	24.2560
16	19.7552	19.8755	-23.4588	23.4716
17	19.7017	19.8531	-22.6867	22.8329
18	19.6688	19.8273	-21.8956	21.9577
19	19.6058	19.7850	-20.9415	20.9773
20	19.4875	19.7290	-20.0486	20.0599
21	19.3816	19.6538	-19.0600	18.8917
22	19.1975	19.5585	-17.9663	17.8725
23	19.0518	19.4317	-16.7615	16.6490
24	18.8028	19.2627	-15.5357	15.3171
25	18.4400	19.0093	-14.1977	13.8807
26	18.0130	18.6423	-12.8371	12.4055
27	17.2961	18.1524	-11.3578	10.7792
28	15.5152	17.4308	-9.7410	9.2858
29	14.7284	16.3760	-7.9840	7.4752
30	12.5507	14.5835	-5.8932	5.6041
31	8.7427	11.6643	-3.8726	3.4926
32	3.5871	7.2291	-1.7185	1.5046
33	0.0294	2.1498	-0.1155	0.0991
34	0.0249	0.5409	-0.0064	0.0072
35	0.0053	0.5255	-0.0060	0.0011
36	0.0088	0.5036	-0.0041	0.0010
37	0.0347	0.5199	0.0013	-0.0001
38	0.0301	0.5014	-0.0000	-0.0013
39	0.0251	0.4653	-0.0001	-0.0039
40	0.0229	0.4697	-0.0007	-0.0030

4.3 Finding the minimum value of the Rastrigin test function

The Rastrigin test function F is defined as

$$F(x1, x2) = x1^2 + x2^2 - 10\cos(2\pi x1) - 10\cos(2\pi x2) + 20$$

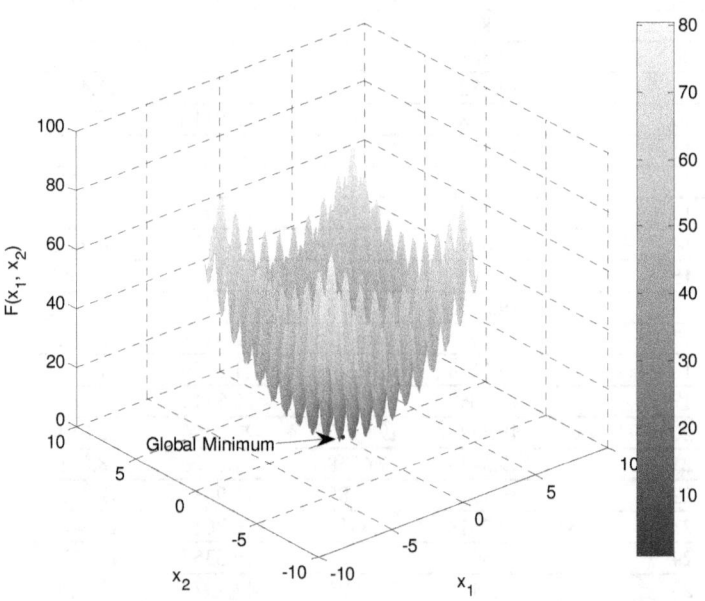

Figure 5. The plot of the Rastrigin test function in 3D.

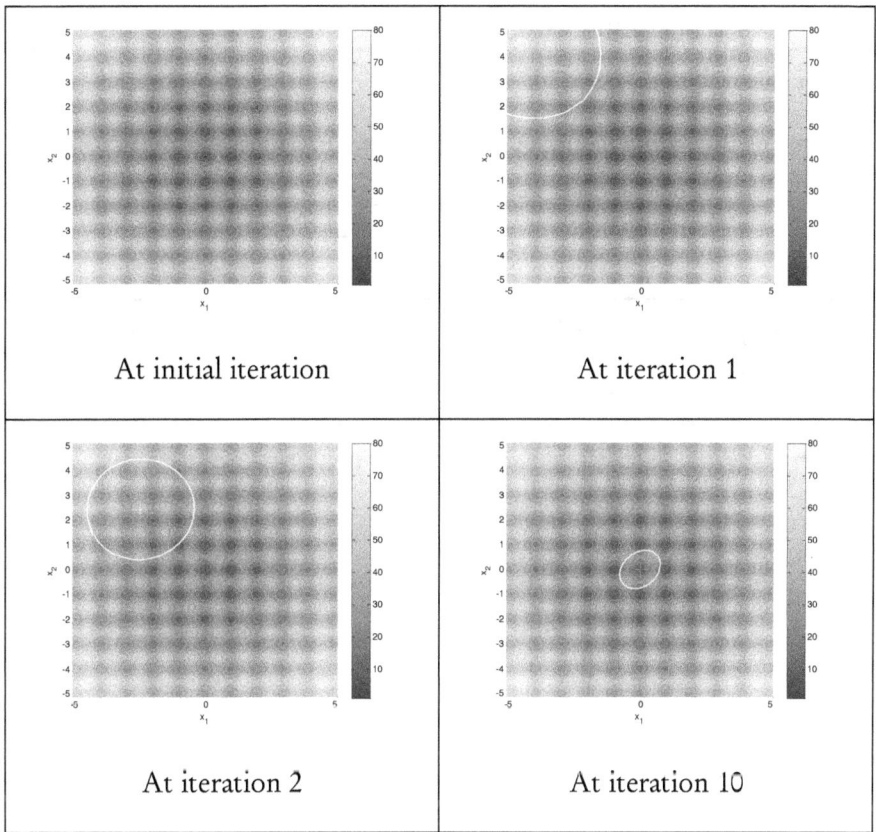

Figure 6. The contour plot of the Rastrigin test function – The search of a global minimum at each iteration.

Table 8. CE algorithm's search for the optimum values

Iteration number	$\hat{\gamma}^{(t)}$	$\hat{v}^{(t)}$	x_1	x_2
1	6.5395	21.4974	-2.4731	2.4337
2	1.6262	7.8246	-1.4369	1.3483
3	0.9443	2.9052	-0.8404	0.8250
4	0.0003	1.6696	-0.4264	0.3515
5	0.0003	1.0270	-0.0684	0.0540
6	0.0018	0.5191	-0.0031	0.0027
7	0.0043	0.3716	0.0007	0.0006
8	0.0021	0.2854	0.0012	-0.0019
9	0.0016	0.2287	0.0009	-0.0020
10	0.0008	0.1880	0.0004	0.0020

4.4 Finding the minimum value of the Hölder table test function

The Hölder table test function f is defined as,

$$F(x1, x2) = -\left| \sin(x1) \cos(x2) e^{\left|1 - \frac{\sqrt{x1^2 + x2^2}}{\pi}\right|} \right|$$

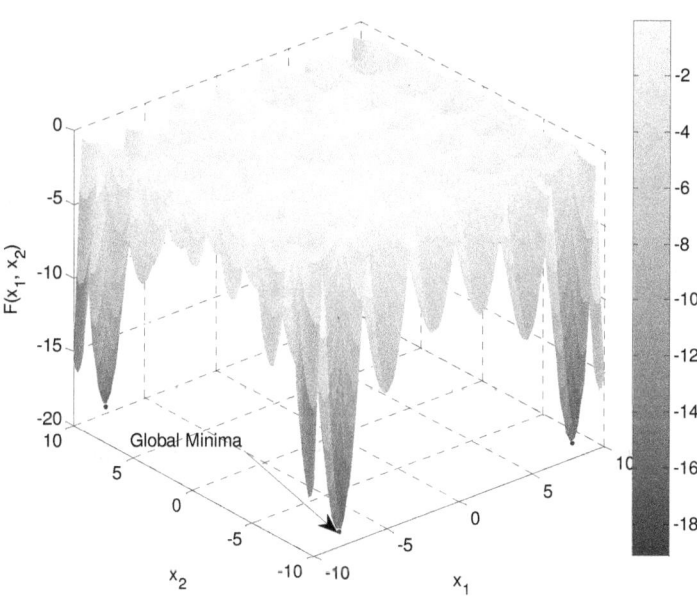

Figure 7. The plot of the Hölder table function in 3D.

Table 9. CE algorithm's search for the optimum values

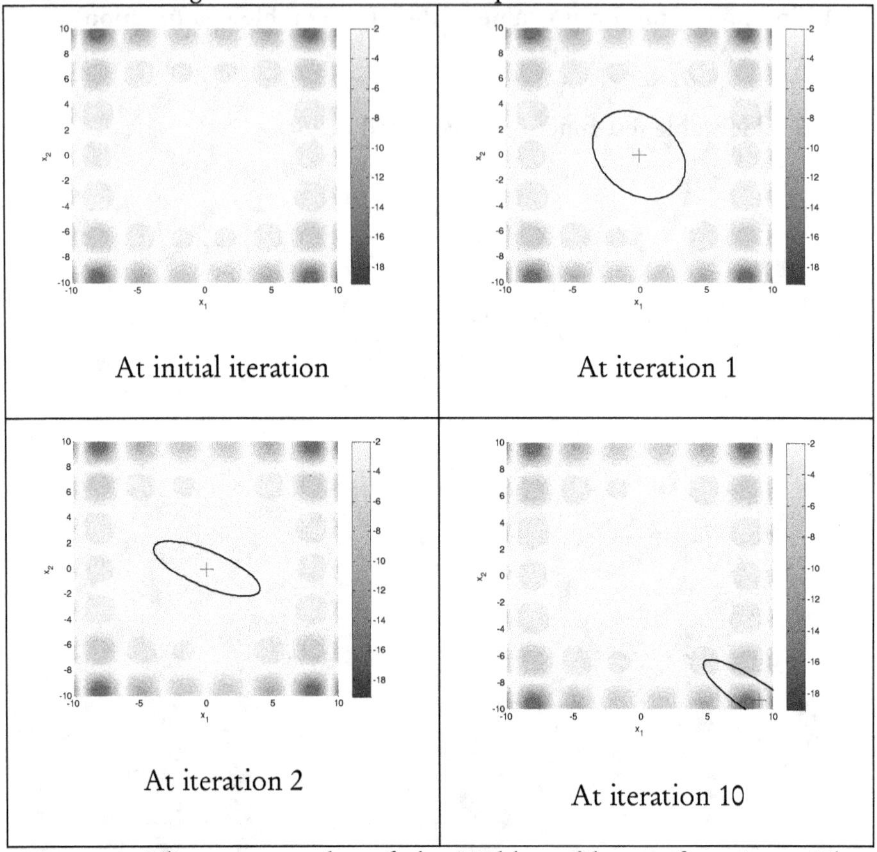

At initial iteration	At iteration 1
At iteration 2	At iteration 10

Figure 8. The contour plot of the Hölder table test function – The search of a global minimum at each iteration.

Table 9. CE algorithm's search for the optimum values

Iteration number	$\hat{\gamma}^{(t)}$	$\hat{v}^{(t)}$	x_1	x_2
1	-2.4783	-1.7217	0.2467	-0.0682
2	-5.1073	-1.7292	1.0025	-0.5533
3	-9.3766	-3.6231	5.8590	-3.2323
4	-9.4566	-7.6462	8.3307	-6.0783
5	-13.8899	-9.3822	8.1698	-6.5472
6	-14.4456	-9.4426	8.2758	-6.7798
7	-14.9241	-9.4609	8.4256	-7.0645
8	-16.2653	-9.4812	9.1023	-8.2246
9	-18.8213	-14.5481	9.0045	-9.3178
10	-19.2075	-18.8066	8.1178	-9.6154

5. Conclusion

This study presented the cross entropy (CE) method for finding minimum/maximum values of some optimization problems. By utilizing the CE method, optimum solutions are achieved for the selected sample optimization problems.

6. References

Abderrahmane, L. H. (2013). Design of a new interleaver using cross entropy method for turbo coding. *Iet Communications*, 7(9), 828-835. doi: 10.1049/iet-com.2012.0599

An, S. G., Yang, S. Y., Ho, S. L., & Ni, P. H. (2012). An Improved Cross-Entropy Method Applied to Inverse Problems. *Ieee Transactions on Magnetics*, 48(2), 327-330. doi: 10.1109/tmag.2011.2173303

Bekker, J. (2013). Multi-Objective Buffer Space Allocation With The Cross-Entropy Method. *International Journal of Simulation Modelling*, 12(1), 50-61. doi: 10.2507/ijsimm12(1)5.228

Bekker, J., & Aldrich, C. (2011). The cross-entropy method in multi-objective optimisation: An assessment. *European Journal of Operational Research*, 211(1), 112-121. doi: 10.1016/j.ejor.2010.10.028

Botev, Z. I., & Kroese, D. P. (2011). The Generalized Cross Entropy Method, with Applications to Probability Density Estimation. *Methodology and Computing in Applied Probability*, 13(1), 1-27. doi: 10.1007/s11009-009-9133-7

Chan, J. C. C., & Kroese, D. P. (2012). Improved cross-entropy method for estimation. *Statistics and Computing*, 22(5), 1031-1040. doi:

10.1007/s11222-011-9275-7

Chen, G., Sarrafzadeh, A., Low, C. P., & Zhang, L. A. (2010). A Self-Organization Mechanism Based on Cross-Entropy Method for P2P-Like Applications. *Acm Transactions on Autonomous and Adaptive Systems*, 5(4). doi: 10.1145/1867713.1867716

Chen, J. C., & Wen, C. K. (2010). Design of two-dimensional zero reference codes with cross-entropy method. *Applied Optics*, 49(18), 3560-3565.

Chen, J. C., Chiu, M. H., Yang, Y. S., & Li, C. P. (2011). A Suboptimal Tone Reservation Algorithm Based on Cross-Entropy Method for PAPR Reduction in OFDM Systems. *Ieee Transactions on Broadcasting*, 57(3), 752-756. doi: 10.1109/tbc.2011.2127590

Chen, J. C., Chiu, M. H., Yang, Y. S., Liao, K. Y., & Li, C. P. (2012). Efficient Capacity-Based Joint Quantized Precoding and Transmit Antenna Selection Using Cross-Entropy Method for Multiuser MIMO Systems. *International Journal of Antennas and Propagation*. doi: 10.1155/2012/965834

Chen, J. C., Wang, S. H., Lee, M. K., & Li, C. P. (2011). Cross-Entropy Method for the Optimization of Optical Alignment Signals With Diffractive Effects. *Journal of Lightwave Technology*, 29(18), 2706-2714. doi: 10.1109/jlt.2011.2163182

Chepuri, K. and T. Homem-de-Mello. (2005). "Solving the Vehicle Routing Problem with Stochastic Demands using the Cross-Entropy Method." *Annals of Operations Research*, 134, 153–181

Cohen, I., B. Golany, and A. Shtub. (2005). "Managing Stochastic Finite Capacity Multi-Project Systems Through the Cross-Entropy Method." *Annals of Operations Research*, 134, 183–199

Connor, J. D. (2008). Antenna array synthesis using the cross entropy method, Ph.D. Thesis, The Florida State University.

Damavandi, M. G., Abbasfar, A., & Michelson, D. G. (2013). Peak Power Reduction of OFDM Systems Through Tone Injection via Parametric Minimum Cross-Entropy Method. *Ieee Transactions on Vehicular Technology*, 62(4), 1838-1843. doi: 10.1109/tvt.2012.2233507

de Boer, P.-T., *et al.* (2005). "A Tutorial on the Cross-Entropy Method." *Annals of Operations Research*, 134(1): 19-67.

Evans, G. E., Sofronov, G. Y., Keith, J. M., & Kroese, D. P. (2011). Estimating change-points in biological sequences via the cross-entropy method. *Annals of Operations Research*, 189(1), 155-165. doi: 10.1007/s10479-010-0687-0

Garvels, M. J. J. (2011). A combined splitting-cross entropy method for rare-event probability estimation of queueing networks. *Annals of Operations Research*, 189(1), 167-185. doi: 10.1007/s10479-009-0608-2

Giunta, A. A. (1997). Aircraft Multidisciplinary Design Optimization Using Design of Experiments Theory and Response Surface Modeling Methods, Ph. D. Thesis, Virginia Polytechnic Institute and State University.

Haber, R. E., del Toro, R. M., & Gajate, A. (2010). Optimal fuzzy control system using the cross-entropy method. A case study of a drilling process. *Information Sciences*, 180(14), 2777-2792. doi: 10.1016/j.ins.2010.03.030

He, D. H., Lee, L. H., Chen, C. H., Fu, M. C., & Wasserkrug, S. (2010). Simulation Optimization Using the Cross-Entropy Method with Optimal Computing Budget Allocation. *Acm Transactions on Modeling and Computer Simulation*, 20(1). doi: 10.1145/1667072.1667076

Ho, S. L., & Yang, S. Y. (2010). The Cross-Entropy Method and Its Application to Inverse Problems. *Ieee Transactions on Magnetics*, 46(8), 3401-3404. doi: 10.1109/tmag.2010.2044380

Ho, S. L., & Yang, S. Y. (2012). Multiobjective Optimization of Inverse Problems Using a Vector Cross Entropy Method. *Ieee Transactions on Magnetics*, 48(2), 247-250. doi: 10.1109/tmag.2011.2175437

Hui, K. P. (2011). Cooperative Cross-Entropy method for generating entangled networks. *Annals of Operations Research*, 189(1), 205-214. doi: 10.1007/s10479-009-0589-1

Kaynar, B., & Ridder, A. (2010). The cross-entropy method with patching for rare-event simulation of large Markov chains. *European Journal of Operational Research*, 207(3), 1380-1397. doi: 10.1016/j.ejor.2010.07.002

Kim, K., Yang, H., & Choi, S. (2014). User Selection Method Adopting Cross-Entropy Method for a Downlink Multiuser MIMO System. *Wireless Personal Communications*, 74(2), 789-802. doi: 10.1007/s11277-013-1321-7

Maher, M., Liu, R. H., & Ngoduy, D. (2013). Signal optimisation using the cross entropy method. *Transportation Research Part C-Emerging Technologies*, 27, 76-88. doi: 10.1016/j.trc.2011.05.018

Margolin, L. (2002). "Application of the Cross-Entropy Method to Scheduling Problems." Master's thesis, Technion, Industrial Engineering.

Margolin, L. (2004). "The Cross-Entropy Method for the Single Machine Total Weighted Tardiness Problem." Unpublished.

Margolin, L. (2005). "On the Convergence of the Cross-Entropy Method." *Annals of Operations Research,* 134(1): 201-214.

Mattrand, C., & Bourinet, J. M. (2014). The cross-entropy method for reliability assessment of cracked structures subjected to random Markovian loads. *Reliability Engineering & System Safety,* 123, 171-182. doi: 10.1016/j.ress.2013.10.009

Minvielle, P., Tantar, E., Tantar, A. A., & Berisset, P. (2011). Sparse Antenna Array Optimization With the Cross-Entropy Method. *Ieee Transactions on Antennas and Propagation,* 59(8), 2862-2871. doi: 10.1109/tap.2011.2158941

Ngoduy, D., & Maher, M. J. (2012). Calibration of second order traffic models using continuous cross entropy method. *Transportation Research Part C-Emerging Technologies,* 24, 102-121. doi: 10.1016/j.trc.2012.02.007

Nguyen, D. M., Thi, H. A. L., & Dinh, T. P. (2014). Solving the Multidimensional Assignment Problem by a Cross-Entropy method. *Journal of Combinatorial Optimization,* 27(4), 808-823. doi: 10.1007/s10878-012-9554-z

Rubinstein, R.Y. & Kroese, D.P. (2004). The Cross-Entropy Method: A Unified Approach to Combinatorial Optimization, Monte-Carlo Simulation and Machine Learning. Springer-Verlag, New York.

Rubinstein, R.Y. & Melamed, B. (1998). Modern simulation and modeling. Wiley series in probability and Statistics.

Rubinstein, R.Y. & Shapiro, A. (1993). Discrete Event Systems: Sensitivity Analysis and Stochastic Optimization via the score function method. Wiley.

Rubinstein, R.Y. (1997). Optimization of computer simulation models with rare events. *European Journal of Operations Research,* pp. 89-112.

Rubinstein, R.Y. (1999). The cross-entropy method for combinatorial and continuous optimization. *Methodology and Computing in Applied Probability,* pp. 127-190.

Rubinstein, R.Y. (2001). Combinatorial optimization, cross-entropy, ants and rare events. In S. Uryasev and P. M. Pardalos, editors, Stochastic Optimization: Algorithms and Applications, Kluwer, pp. 304-358.

Rubinstein, R.Y. (2002). "Cross-Entropy and Rare-Events for Maximal Cut and Bipartition Problems." *ACM Transactions on Modeling and Computer Simulation,* 27–53

Salama, M. S., & Shen, F. (2010). Stochastic inversion of ocean color data using the cross-entropy method. *Optics Express,* 18(2), 479-499.

Selvakumar, A. I. (2011). Enhanced cross-entropy method for dynamic economic dispatch with valve-point effects. *International Journal of Electrical Power & Energy Systems*, 33(3), 783-790. doi: 10.1016/j.ijepes.2011.01.001

Selvan, S. E., Subathra, M. S. P., Christinal, A. H., & Amato, U. (2013). On the benefits of Laplace samples in solving a rare event problem using cross-entropy method. *Applied Mathematics and Computation*, 225, 843-859. doi: 10.1016/j.amc.2013.10.011

Wang, Q., Wang, H. X., & Yan, Y. (2011). Fast reconstruction of computerized tomography images based on the cross-entropy method. *Flow Measurement and Instrumentation*, 22(4), 295-302. doi: 10.1016/j.flowmeasinst.2011.03.010

Weatherspoon, M. H., Connor, J. D., & Foo, S. Y. (2013). Shaped beam synthesis of phased arrays using the cross entropy method. *International Journal of Numerical Modelling-Electronic Networks Devices and Fields*, 26(6), 630-642. doi: 10.1002/jnm.1893

Yildiz, T., & Yercan, F. (2010). The Cross-Entropy Method for Combinatorial Optimization Problems Of Seaport Logistics Terminal. *Transport*, 25(4), 411-422. doi: 10.3846/transport.2010.51

Zhang, W., Yang, S. Y., Bai, Y. A., & Machado, J. M. (2010). The cross-entropy method and its application to minimize the ripple of magnetic levitation forces of a maglev system. *International Journal of Applied Electromagnetics and Mechanics*, 33(3-4), 1063-1068. doi: 10.3233/jae-2010-1221

Chapter 5. Logistics systems and optimization strategies under uncertain operational environment

Abstract. Transportation and logistics systems are characterized by their highly dynamic structures along with numerous interconnected processes. The natures of these systems involve various levels of resource allocation decisions where usually it is not always possible to execute these decisions in the field on time at the best possible way because of the unpredictable factors in plans. By considering the uncertain operational environment, this study explores the uncertainty issue within operational systems and deals with the problem of allocating resources to maximize expected total profit and minimize inefficiencies under uncertainty. The aim is to design, develop, visualize, and effectively deal with a more realistic model to satisfy uncertain demand nodes by leaving minimal or none unsatisfied zones within an operational environment at seaports, transportation, logistics and supply chain systems. A representative optimization model, which is developed to address the uncertainty issue, has been solved by using an optimization algorithm. The results show that operational plans without the utilization of uncertainty models could have negative impacts, including increased emissions, negative environmental effects, along with higher costs to organizations.

Keywords: Uncertainty, optimization algorithm, resource allocation problem, transportation, logistics, environment

1. Introduction

Logistics systems have been at the center of attention of enterprises for over a decade with today's heightened expectation of customers and harsh competitive environment in markets. Enterprises are continuously seeking the best development and enterprise-level best solution strategies for their logistics systems to meet the present and possible future expectations and to stay beyond their counterparts.

With the advancement of information, communication systems, and optimization techniques, it became possible to model and to obtain solutions for large-scale complex problems.

Optimization is a well-established field due to comprehensive research conducted over past decades. In decision science, optimization is an essential tool. As stated by Nocedal (1999), optimization is an important tool in decision science and in the analysis of physical systems. In order to benefit from optimization, it is necessary to identify some objectives and then some quantitative measures of the system.

The objective of a system could be maximizing profit, minimizing total time for completing a task or a group of tasks, or maximizing/minimizing any quantity or composition of quantities that can be represented by mathematical models. The objective depends on certain characteristics of the system, called variables or unknowns. The goal of optimization is to find correct values for the variables that optimize the objective. The variables are usually restricted, or constrained, in some way (Nocedal 1999). Many types of optimization problems have been addressed and various types of algorithms have been researched. Several methodologies for optimizing objectives of systems have been used in various practical applications and the range of applications is continuously growing (Arora 2007).

For example, the objective function might be linear or nonlinear, differentiable or nondifferentiable, concave or convex, etc. The decision variables might be continuous or discrete. The feasible region might be convex or nonconvex. These differences each impact how the model can be solved, and thus optimization models are classified according to these differences (Sarker 2002).

Global optimization is made even more difficult because supply chains need to be designed for and operated in uncertain environments (Simchi-Levi 2004). Operations research uses quantitative models to analyze and predict the behavior of systems, and to provide information for decision makers. Two key concepts in operations research are optimization and uncertainty. Uncertainty is emphasized in operations research that could be called "stochastic operations research" in which uncertainty is described by stochastic models. The typical models in stochastic operations research are queuing models, inventory models, financial engineering models, reliability models, and simulation models (Dohi 2009).

This paper is organized as follows. Section 2 reviews the literature. Section 3 provides general background information on mathematical programming with uncertainty components and recourse variables. The study is concluded in Section 4.

2. Literature review

Hall (2003) stated that transportation science covers research from many fields such as geography, economics, and location theory. Methodologies of transportation science come from physics, operations research, probability, and control theory; it is fundamentally a quantitative discipline, relying on mathematical models and optimization algorithms to explain the phenomena of transportation (Hall 2003). Frazelle (2002) explained the overall goal of transportation as it should be: to connect sourcing locations with customers at the lowest possible transportation cost within the constraints of the customer service policy. Thus, the transportation optimization equation could be expressed as follows: minimizing total transportation costs subject to customer service policy constraints (Frazelle 2002). Frazelle (2002) also emphasized that transportation expenses are rising quickly versus other logistics costs, with smaller, more frequent orders, increasing international trade and global logistics, rising fuel charges, labor shortages, decreased carrier competition due to carrier mergers and acquisitions, and increased union penetration in the labor market.

Additionally, Simchi-Levi (2004) acknowledged that it is challenging to design and operate a supply chain so that total system wide costs are minimized and system wide service levels are maintained. Indeed, it is frequently difficult to operate a single facility so that costs are minimized and service level is maintained. The difficulty increases exponentially when an entire system is being considered. The process of finding the best system wide strategy is known as global optimization (Simchi-Levi 2004). The process of identifying objectives, variables, and constraints for a given problem is known as modeling.

Construction of an appropriate model is the first step— sometimes the most important step— in the optimization process. If the model is too simplistic, it will not give useful insights into the practical problem, but if it is too complex, it may become too difficult to solve (Nocedal 1999). Before solving an optimization model, it is important to consider the form and mathematical properties of the objective function, constraints, and decision variables.

There are a large number of optimization problems in organized systems, such as in industrial systems, business systems, transportation and logistics systems, where at strategic, tactical, and operational levels, planners, analysts, strategists, and engineers are confronted with uncertainty. Many optimization problems arising from these systems have deterministic (certain) parts along with uncertain components, which planners could disambiguated based on the predictable and probabilistic information. Along with the other various factors, Pardalos (2004) pointed out the crucial issue of developing efficient methods of analyzing this information in order to understand the internal structure of the market and make effective strategic decisions for the successful operation of a business.

In addition, referring to the efficiency of the transportation infrastructure, Hall (2003) expressed that planners and engineers need to forecast the demand of transportation to make informed transportation infrastructure planning decisions. Arora (2007) stated for any engineering system that the uncertainties in system characteristics and demand prevent assurance from being given with absolute certainty. For supply chain systems, Simchi-Levi (2004) emphasized that it is necessary to design systems that eliminate as much uncertainty as possible and it is necessary to deal effectively with the uncertainty that remains.

Much research has been devoted to tackling optimization problems under uncertainty. Researchers have investigated various systems and have dealt with the uncertainty issue. Table 1 highlights recent and leading studies on the applications of stochastic programming with uncertainty depending on the fields studied.

Table 1a. Studies on stochastic programming with uncertainty (Optimization problems).

A multi-objective robust stochastic programming model for disaster relief logistics under uncertainty (Bozorgi-Amiri *et al.* 2013).
A multistage stochastic programming approach for capital budgeting problems under uncertainty (Beraldi *et al.* 2013).
Integration of Scheduling and Dynamic Optimization of Batch Processes under Uncertainty: Two-Stage Stochastic Programming Approach and Enhanced Generalized Benders Decomposition Algorithm (Chu and You 2013).
Scheduling jobs sharing multiple resources under uncertainty: A stochastic programming approach (Keller and Bayraksan 2010).
Determining supply requirement in the sales-and-operations-planning (S&OP) process under demand uncertainty: a stochastic programming formulation and a spreadsheet implementation (Sodhi and Tang 2011).
A multi-stage stochastic programming approach for production planning with uncertainty in the quality of raw materials and demand (Zanjani *et al.* 2010).
Capacities-based supply chain network design considering demand uncertainty using two-stage stochastic programming (Singh *et al.* 2013).
Location of cross-docking centers and vehicle routing scheduling under uncertainty: A fuzzy possibilistic-stochastic programming model (Mousavi *et al.* 2014).
A stochastic programming winner determination model for truckload procurement under shipment uncertainty (Ma *et al.* 2010).
Stochastic programming approach to re-designing a warehouse network under uncertainty (Kiya and Davoudpour 2012).
Solution strategies for multistage stochastic programming with endogenous uncertainties (Gupta and Grossmann 2011).
A stochastic programming approach for optimal microgrid economic operation under uncertainty using $2m+1$ point estimate method (Mohammadi *et al.* 2013).

Table 1b. Studies on stochastic programming with uncertainty (Industry).

Design under uncertainty of hydrocarbon biorefinery supply chains: Multiobjective stochastic programming models, decomposition algorithm, and a Comparison between CVaR and downside risk (Gebreslassie *et al.* 2012).
Electric sector investments under technological and policy-related uncertainties: a stochastic programming approach (Bistline and Weyant 2013).
Sustainable development and planning of coal industry under uncertainty using system dynamic and stochastic programming (Xu and Wu 2010).

Table 1c. Studies on stochastic programming with uncertainty (Environment).

An interval-fuzzy two-stage stochastic programming model for planning carbon dioxide trading under uncertainty (Li *et al.* 2011).
Energy and environmental systems planning under uncertainty-An inexact fuzzy-stochastic programming approach (Li *et al.* 2010).
Two-Stage Stochastic Programming Model for Planning CO_2 Utilization and Disposal Infrastructure Considering the Uncertainty in the CO_2 Emission (Han and Lee 2011).
Developing a Two-Stage Stochastic Programming Model for CO_2 Disposal Planning under Uncertainty (Han *et al.* 2012).
A two-stage inexact-stochastic programming model for planning carbon dioxide emission trading under uncertainty (Chen *et al.* 2010).

Table 1d. Studies on stochastic programming with uncertainty (water resources management).

Inexact Fuzzy-Stochastic Programming for Water Resources Management Under Multiple Uncertainties (Guo *et al.* 2010).
Inexact joint-probabilistic stochastic programming for water resources management under uncertainty (Li and Huang 2010).
An inexact fuzzy parameter two-stage stochastic programming model for irrigation water allocation under uncertainty (Li *et al.* 2013).

Table 1e. Studies on stochastic programming with uncertainty (Waste management).

An interval-parameter mean-CVaR two-stage stochastic programming approach for waste management under uncertainty (Dai *et al.* 2014).
A Superiority-Inferiority-Based Inexact Fuzzy Stochastic Programming Approach for Solid Waste Management Under Uncertainty (Tan *et al.* 2010).

Table 1f. Studies on stochastic programming with uncertainty (Energy systems).

Development of an interval multi-stage stochastic programming model for regional energy systems planning and GHG emission control under uncertainty (Li *et al.* 2012).
Decomposition Based Stochastic Programming Approach for Polygeneration Energy Systems Design under Uncertainty (Liu *et al.* 2010).
An interval fixed-mix stochastic programming method for greenhouse gas mitigation in energy systems under uncertainty (Xie *et al.* 2010).

Moreover, while taking into consideration efficient systems design under uncertainty, critical awareness about the environment is important. The transport sector is one of the few sectors that have ever-increasing greenhouse gas (GHG) emissions. According to Intergovernmental Panel on Climate Change (IPCC) reports, transport accounts for 13.1% of GHG emissions, 23% of global carbon dioxide

(CO_2) emissions, 26% of total world energy use, and the transport sector is one of the constantly growing sectors. Thus, it is also critical to take seriously into consideration inefficient transportation and logistics system designs that can have negative effects on the environment.

A report by Pronello and Andrè (2000) proposed that pollution caused by a set of vehicle routes can be unreliable. However, it is obvious that reducing the total distance and thus putting to practical use of optimal routes by algorithms will provide environmental benefits as a result of the reduction in fuel consumption and the consequent pollutants (Sbihi and Eglese 2007). Palmer (2004) also analyzed a study about the connection between efficient network design and thus reducing environmental pollutants.

3. A Background: Mathematical programming

This section presents a brief general background about mathematical programming with *second-stage* or *recourse* variables. As stated by Sahinidis (2004), second-stage variables can be interpreted as corrective measures or recourse against any infeasibilities arising due to a particular realization of uncertainty. The most common programming models are stochastic linear/non-linear/integer programming, probabilistic or chance-constrained programming, robust stochastic programming, stochastic dynamic programming, and fuzzy programming. The next section will present brief fundamentals of mathematical programming with uncertainty. Further details of mathematical programming concepts have been well documented in the textbooks of Bertsekas and Tsitsiklis (1996), Birge and Louveaux (1997), Kall and Wallace (1994), Prékopa (1995), and Zimmermann (1991). Early works involving uncertainty studies were included research by Beale (1955), Bellman (1957), Bellman and Zadeh (1970), Charnes and Cooper (1959), Dantzig (1955), and Tintner (1955).

3.1 Stochastic linear programming with recourse

A stochastic programming can be stated as

$$\min\ f(X) = C^t X = \sum_{j=1}^{n} c_j x_j$$

subject to

$$A_i^T X = \sum_{j=1}^{n} a_{ij} x_j \leq b,\ i = 1, 2, 3, \ldots, m$$

$$x_j \geq 0,\ j = 1, 2, 3, \ldots, n$$

Based on the above formulation, the general formulation for two-stage (*recourse*) stochastic linear program (Birge and Louveaux 1997; Infanger 1994; Kall and Wallace 1994 and Prékopa 1995) is

The first stage of the problem,

$$\min\ c^t x + E_{\omega \in \Omega}[Q(x, \omega)]$$

$$s.t.\ Ax \leq b,\ x \geq 0,\ x \in X$$

and the second stage of the problem,

$$Q(x, \omega) = \min f(\omega)^t y$$

Subject to

$$D(\omega) y \geq h(\omega) + T(\omega) x,\ y \in Y$$

ω is a random variable from a probability space (Ω, F, P) and within the interval $\Omega = \left[\omega^{L.B.}, \omega^{U.B.}\right]$. The expected value function $E_{\omega \in \Omega}[Q(x, \omega)]$ is

$$E_{\omega \in \Omega}[Q(x, \omega)] = \int_{\omega^{L.B.}}^{\omega^{U.B.}} Q(x, \omega) d(P)$$

The first stage decision variables are represented by vector x and the second stage variables are represented by vector y.

A is a first stage $(m \times n)$ constraint matrix. b is a first stage (m component) right hand side vector. c is a first stage (n component) cost vector. x is a vector of first stage vector variables. y is a vector of second stage variables. Ω is the all possible variables of the random variable ω. D is a second stage recourse matrix. $T(\omega)$ is a second stage matrix. $h(\omega)$ is a second stage right hand side vector. $Q(\cdot)$ is a cost function of the second stage. P is the probability measure of the variable ω. The variables $\omega^{U.B.}, \omega^{L.B.}$ are upper and lower bounds of the variable ω.

Probabilistic/Chance-constrained programming:

$$\min\ c'x$$
$$s.t.\ P(Tx \geq \xi) \geq p$$
$$Ax = b,\ x \geq 0$$

p is the fixed probability. $P(Tx \geq \xi) \geq p$ is the probabilistic constraint.

Stochastic non-linear programming with recourse:

The first stage of the problem,

$$\min\ f(x) + E_{\omega \in \Omega}[Q(x, \omega)]$$

subject to

$$g(x) \leq 0$$

with the second stage of the problem,

$$Q(x, \omega) = \min F(\omega, x, y)$$

subject to

$$G(\omega, x, y) \leq 0,\ y \in Y$$

where ω is a random variable from a probability space (Ω, F, P).

3.2 A sample model with an uncertainty component

The expected cost φ is the total cost and plus the expected revenue lost due to the unassigned cargo spaces/seats.

The objective functions $\varphi_{1,2}$ are

$$\text{Min. } \varphi_1 = \sum_{i=1}^{\mu}\sum_{j=1}^{v} c_{ij}x_{ij} + \left[\sum_{j=1}^{v} k_j \omega_j - \sum_{j=1}^{v} k_j \sum_{h=1}^{\rho} \gamma_{hj}\psi_{hj}\right]$$

$$\text{Min. } \varphi_2 = \sum_{i=1}^{\mu}\sum_{j=1}^{v} c_{ij}x_{ij} + \left[\sum_{jh} k_j \lambda_{jh}\beta_{jh}\right]$$

subject to:

$$\psi_{jh} \leq \sum_{i}\pi_{ij}x_{ij}$$

$$\beta_{jh} = \delta_{jh} - \psi_{jh}$$

Vehicle balance:

$$\sum_{j} x_{ij} \leq v_i$$

Demand balance:

$$\sum_{i}\pi_{ij}x_{ij} \geq \sum_{h\$\Delta b_{jh}} \psi_{jh}$$

where

$$i, j, h \in \mathfrak{R}_1$$

$i = 1, 2, 3, ..., \mu - 1$ is vehicle types and unassigned cargo container/passenger. $i = \mu$, if unassigned. $j = 1, 2, 3, ..., v - 1$ is the type of route to which a vehicle is assigned. $j = v$, if unassigned. $h = 1, 2, 3, ..., \rho$ is the demand state. c_{ij} is the cost per vehicle type i on route j. π_{ij} is the capacity of vehicle type i on route j. v_i is the vehicle availability. λ_{jh} is the probability of demand state h on route j. δ_{jh} is the demand distribution on route j based on the demand state h. γ_{hj} is the probability of exceeding demand increment h on route j. ω_j is the expected demand on route j. k_j is the revenue lost. x_{ij} is the vehicle type i assigned to route j. ψ_{hj} is the load actually carried on route j based on the demand state h.

β_{hj} is the load not carried on route j based on the demand state h. $\sum_j x_{ij}$ is the total available vehicle type i on route j. The first term of the objective function $\sum_{i=1}^{\mu} \sum_{j=1}^{v} c_{ij} x_{ij}$ is the operating cost. $\sum_{j=1}^{v} k_j$ is the total revenue lost and $\sum_{j=1}^{v} \omega_j$ is the total expected demand, thus $\sum_{j=1}^{v} k_j \omega_j$ is the expected revenue, in case all allocations are made sufficiently. The second terms of the objective functions $\left[\sum_{j=1}^{v} k_j \omega_j - \sum_{j=1}^{v} k_j \sum_{h=1}^{\rho} \gamma_{hj} \psi_{hj} \right]$ and $\left[\sum_{jh} k_j \lambda_{jh} \beta_{jh} \right]$ are the expected revenue lost due to unassigned spaces. For more information about the mathematical model, refer to a slightly different model from Dantzig (1963).

3.3 Initialization with data and simulation

This sub-section initializes the sample model described in the previous section with data to represent a realistic problem. Sample data initialization includes few steps as shown in Figure 1. To solve the optimization problem with the initialized data, IpOpt algorithm is used. The IpOpt algorithm implements an interior point line search filter method that aims to find a local solution of the objective function subject to constraints. The mathematical details of the algorithm are in various publications (Waechter 2002; Waechter and Biegler 2005, 2006; Nocedal *et al.* 2005). The algorithm is intended to solve non-linear problems; however, it is also capable of providing solutions for various other types of problems as well as linear problems.

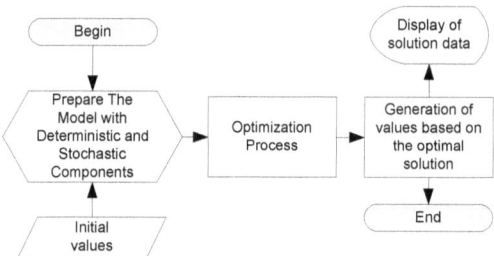

Figure 1. The flowchart of the optimization process

Waechter and Biegler (2006) stated that the interest in efficient optimization methods led to the development of interior-point or barrier methods for large-scale nonlinear programming. In addition, Waechter and Biegler (2006) further emphasized the fact that in particular, these methods provide an attractive alternative to active set strategies in handling problems with large numbers of inequality constraints. For the convergence properties of interior-point methods (Forsgren 2002) over the past years there has been a better understanding and efficient algorithms with desirable global and local convergence properties have been developed (Waechter and Biegler 2006). Waechter and Biegler (2006) also stated that to allow convergence from poor starting points, interior-point methods, in both trust region and line-search frameworks, have been developed that use exact penalty merit functions to enforce progress toward the solution (Byrd *et al.* 2000; Tits *et al.* 2003; Yamashita 1998; Waechter and Biegler

2006).

The next part shows a solution to the optimization problem under uncertainty having 100 vehicles and 25 different routes. In this case, the total number of variables is 3250 (See Table 2). Table 2 shows the optimization process.

Table 2. The optimization process (IpOpt algorithm)

Iter. No.	Objective	Inf_pr	Inf_du	Lg(mu)	||d||	Lg(rg)	Alpha_du	Alpha_pr	Ls
0	2.0000000e-002	7.64e+004	1.19e+000	0.0	0.00e+000	-	0.00e+000	0.00e+000	0
1	2.2453154e-002	7.32e+004	2.83e+001	-5.3	2.74e+001	-	1.40e-003	4.29e-002h	1
2	3.5510215e-002	6.59e+004	2.61e+001	-1.2	1.94e+002	-	1.66e-003	9.94e-002h	1
3	3.4950269e-002	6.42e+004	2.54e+001	-5.3	1.35e+003	-	1.41e-002	2.62e-002h	1
4	4.0035380e-002	6.19e+004	2.45e+001	-1.2	6.87e+002	-	6.64e-002	3.45e-002h	1
5	3.6059254e-002	4.93e+004	1.95e+001	-5.3	6.43e+002	-	1.70e-002	2.05e-001h	1
6	3.5044461e-002	4.36e+004	1.72e+001	-5.3	5.15e+002	-	8.76e-002	1.15e-001h	1
7	3.5879943e-002	3.91e+004	1.55e+001	-5.4	4.64e+002	-	5.95e-002	1.03e-001h	1
8	4.0438994e-002	3.58e+004	1.42e+001	-5.4	4.33e+002	-	1.07e-001	8.31e-002h	1
9	7.0272037e-002	3.20e+004	1.27e+001	-5.5	4.23e+002	-	4.96e-002	1.07e-001h	1
10	2.2892969e-001	2.94e+004	1.35e+001	-5.5	3.94e+002	-	3.01e-002	8.25e-002h	1
11	1.1670179e+002	8.78e+003	2.11e+001	-1.6	2.98e+002	-	1.44e-003	7.01e-001f	1
12	1.0487235e+002	2.12e+003	4.99e+000	-5.5	9.57e+001	-	1.36e-002	7.59e-001h	1
13	9.4358075e+001	5.93e+002	1.36e+000	-5.5	2.74e+001	-	3.91e-001	7.20e-001h	1
14	7.7650402e+001	8.49e+001	2.20e-001	-5.7	1.95e+001	-	4.39e-001	8.57e-001h	1
15	5.4069763e+001	2.51e+000	5.78e-003	-6.0	2.43e+001	-	9.80e-001	9.70e-001f	1
16	2.7939077e+000	2.10e-002	9.09e-005	-7.4	5.17e+001	-	9.79e-001	9.92e-001f	1
17	3.0061904e-002	2.06e-004	9.32e-007	-8.9	2.79e+000	-	9.90e-001	9.90e-001f	1
18	4.8234800e-005	3.42e-007	1.50e-009	-10.9	3.01e-002	-	9.98e-001	9.98e-001f	1
19	3.8114012e-008	7.56e-009	4.81e-005	-9.5	6.72e+000	-	1.00e+000	9.79e-001h	1
20	8.4073551e-011	6.92e-009	3.43e+000	-9.5	1.09e-005	-4.0	1.00e+000	3.86e-002h	1
21	4.3991391e-008	1.71e-009	1.18e+000	-9.3	4.98e-005	-4.5	1.00e+000	7.75e-001h	1
22	2.4409058e-008	7.95e-010	4.02e-001	-9.3	1.49e-004	-5.0	1.00e+000	5.03e-001f	1
23	1.8460108e-009	1.35e-010	6.91e-011	-9.3	6.08e+000	-	1.00e+000	1.00e+000f	1

Table 2 and Figure 2 show the optimization process. Table 3 and Table 4 show the final output of the process. The column "*iter*" is the iteration counter and the column "*objective*" is the current value of the objective function. Other values of the columns are primal infeasibility (*inf_pr*), dual infeasibility (*inf_du*), logarithm of current barrier parameter (*lg(mu)*), max-norm of the primal search direction (||*d*||), logarithm of Hessian perturbation (*lg(rg)*), dual step size (*alpha_du*), primal step size (*alpha_pr*), and number of backtracking steps (*ls*).

Table 3. About the optimized model.

Number of nonzeros in equality constraint Jacobian	3218
Number of nonzeros in inequality constraint Jacobian	5709
Number of nonzeros in Lagrangian Hessian	0
Total number of variables	3250
variables with only lower bounds	3250
variables with lower and upper bounds	0
variables with only upper bounds	0
Total number of equality constraints	2
Total number of inequality constraints	125
inequality constraints with only lower bounds	25
inequality constraints with lower and upper bounds	0
inequality constraints with only upper bounds	100

Ipopt shows information (see Tables 2, 3, 4) of the optimization procedure and closes with statistical information about the computational effort. In the Table 2, the first column "*iter*" is the iteration counter. It is not reset after an update of the barrier parameter or when the algorithm switched between the restoration phase and the regular algorithm. The next two columns "*objective*" and "*inf_pr*" indicate the value of the objective function. The fourth column "*inf_du*" is a measure of optimality; as the Ipopt aims to find a point satisfying the optimality conditions. The last column "*ls*" is an indication about how many trial points needed to be evaluated. Additionally, as the value of the barrier parameter is going to zero, there is a decrease in the number in column "*lg(mu)*". Furthermore, the larger the step sizes of columns "*alpha_du*" and "*alpha_pr*", the better is the usually the progress. (For more details about the algorithm, refer to https://projects.coin-or.org/Ipopt).

Table 4. The final output of the optimization process.

	(unscaled)	(scaled)
Objective	1.8460108023708330e-009	1.8460108023708330e-009
Dual infeasibility	6.9146319266848223e-011	6.9146319266848223e-011
Constraint violation	1.3451350211095411e-010	1.6007106751203540e-010
Complementarity	7.5684299981862521e-010	7.5684299981862521e-010
Overall NLP error	7.5684299981862521e-010	7.5684299981862521e-010

The process provides an optimal solution for the sample model having over a hundred thousand variables/constraints (See Table 2). The return of an optimal solution value takes less than few minutes on a computer having an Intel® Core™2 Quad Q6600 at 2.4GHz CPU. Many commercial optimization algorithms (i.e. IBM® Cplex, and more) return solutions in less than a few seconds even with millions of variables and constraints.

4. Conclusion

While there is no single effective solution that can be applied to various other types of real-world problems having some uncertainty components, the IpOpt optimization method presented on this paper and theoretically applied based on a typical scenario. In brief, this paper has introduced optimization issues under uncertainty along with brief fundamentals of mathematical programming concepts that address various possible uncertainty situations. It then presented a sample optimization model with uncertainty variables. Finally, it showed optimal solutions found at the last iteration of the optimization algorithm. The optimal solution found at the last iteration provides critical information, which addresses the best system configuration and conditions under changing operational environment.

The sample optimization model put the IpOpt optimization algorithm to practical use; this model was developed to address the uncertainty issue. The results indicate that operational plans without taking into consideration of uncertainty models could have negative impacts that lead to non-optimal operational design and thus, higher costs to organizations along with possible harmful environmental effects (i.e. increased emissions, pollutants, and more). This study solved a sample model having more than a hundred thousand variables/constraints. The aim was to visualize an optimal solution under uncertainty for a sample logistics problem. Many other solvers and optimization algorithms are capable of solving highly complex problems with millions of variables in a short period depending on the CPU power of the computer.

5. References

Arora, J.S. (Ed.). (2007) Optimization of Structural and Mechanical Systems. River Edge, NJ, USA: World Scientific. p 1-271.

Beale, E.M.L. (1955) On minimizing a convex function subject to linear inequalities. *Journal of the Royal Statistical Society*, 17B, pp. 173-184.

Bellman, R. E. (1957) Dynamic programming. Princeton, PA: Princeton University Press.

Bellman, R., Zadeh, L.A. (1970) Decision-making in a fuzzy environment. *Management Science*, 17, pp. 141-161.

Beraldi, P., Violi, A., De Simone, F., Costabile, M., Massabo, I., & Russo, E. (2013). A multistage stochastic programming approach for capital budgeting problems under uncertainty. *Ima Journal of Management Mathematics*, 24(1), 89-110. doi: 10.1093/imaman/dps018

Bertsekas, D. P., Tsitsiklis, J. N. (1996) Neuro-dynamic programming. Belmont, MA: Athena Scientific.

Birge, J. R., Louveaux, F. V. (1997) Introduction to stochastic programming. New York, NY: Springer.

Bistline, J. E., & Weyant, J. P. (2013). Electric sector investments under technological and policy-related uncertainties: a stochastic programming approach. *Climatic Change*, 121(2), 143-160. doi: 10.1007/s10584-013-0859-4

Bozorgi-Amiri, A., Jabalameli, M. S., & Al-e-Hashem, S. (2013). A multi-objective robust stochastic programming model for disaster relief logistics under uncertainty. *Or Spectrum*, 35(4), 905-933. doi: 10.1007/s00291-011-0268-x

Byrd, R. H., Gilbert, J. Ch., Nocedal, J. (2000) A trust region method based on interior point techniques for nonlinear programming. *Mathematical Programming*, 89, 149-185

Charnes, A., Cooper, W.W. (1959) Chance-constrained programming. *Management Science*, 6, pp. 73-79.

Chen, W. T., Li, Y. P., Huang, G. H., Chen, X., & Li, Y. F. (2010). A two-stage inexact-stochastic programming model for planning carbon dioxide emission trading under uncertainty. *Applied Energy*, 87(3), 1033-1047. doi: 10.1016/j.apenergy.2009.09.016

Chu, Y. F., & You, F. Q. (2013). Integration of Scheduling and Dynamic Optimization of Batch Processes under Uncertainty: Two-Stage Stochastic Programming Approach and Enhanced Generalized Benders Decomposition Algorithm. *Industrial & Engineering Chemistry Research*, 52(47), 16851-16869. doi: 10.1021/ie402621t

Dai, C., Cai, X. H., Cai, Y. P., Huo, Q., Lv, Y., & Huang, G. H. (2014). An interval-parameter mean-CVaR two-stage stochastic programming approach for waste management under uncertainty. *Stochastic Environmental Research and Risk Assessment*, 28(2), 167-187. doi: 10.1007/s00477-013-0738-6

Dantzig, G.B. (1955) Linear programming under uncertainty. Management Science 1, pp. 197–206.

Dantzig, G.B. (1963) Chapter 28. In Linear Programming and Extensions. Princeton University Press, Princeton, New Jersey.

Dohi, T. (Ed.). (2009) Recent Advances in Stochastic Operations Rsearch II. SGP: World Scientific, p 6.

Forsgren, A., Gill, P. E., Wright, M. H. (2002) Interior methods for nonlinear optimization. *SIAM Review*, 44 (4), 525–597

Frazelle, E. H. (2002) Supply Chain Strategy.Blacklick, OH, USA: McGraw-Hill Education Group, 2002. p 169-174.

Gebreslassie, B. H., Yao, Y., & You, F. Q. (2012). Design under uncertainty of hydrocarbon biorefinery supply chains: Multiobjective stochastic programming models, decomposition algorithm, and a Comparison between CVaR and downside risk. *Aiche Journal*, 58(7), 2155-2179. doi: 10.1002/aic.13844

Guo, P., Huang, G. H., & Li, Y. P. (2010). Inexact Fuzzy-Stochastic Programming for Water Resources Management Under Multiple Uncertainties. *Environmental Modeling & Assessment*, 15(2), 111-124. doi: 10.1007/s10666-009-9194-6

Gupta, V., & Grossmann, I. E. (2011). Solution strategies for multistage stochastic programming with endogenous uncertainties. *Computers & Chemical Engineering*, 35(11), 2235-2247. doi: 10.1016/j.compchemeng.2010.11.013

Hall, R.W.(Ed.). (2003) Handbook of Transportation Science, Second Edition. Secaucus, NJ, USA: Kluwer Academic Publishers. p 2-39.

Han, J. H., & Lee, I. B. (2011). Two-Stage Stochastic Programming

Model for Planning CO2 Utilization and Disposal Infrastructure Considering the Uncertainty in the CO2 Emission. *Industrial & Engineering Chemistry Research*, 50(23), 13435-13443. doi: 10.1021/ie200362y

Han, J. H., Ryu, J. H., & Lee, I. B. (2012). Developing a Two-Stage Stochastic Programming Model for CO2 Disposal Planning under Uncertainty. *Industrial & Engineering Chemistry Research*, 51(8), 3368-3380. doi: 10.1021/ie201148x

Infanger, G. (1994) Planning under uncertainty: Solving large scale stochastic linear programs. Danvers, MA: Boyd and Fraser Publishing Co.

Kall, P., Wallace, S. W. (1994) Stochastic programming. New York, NY: Wiley.

Keller, B., & Bayraksan, G. N. (2010). Scheduling jobs sharing multiple resources under uncertainty: A stochastic programming approach. *Iie Transactions*, 42(1), 16-30. doi: 10.1080/07408170902942683

Kiya, F., & Davoudpour, H. (2012). Stochastic programming approach to re-designing a warehouse network under uncertainty. *Transportation Research Part E-Logistics and Transportation Review*, 48(5), 919-936. doi: 10.1016/j.tre.2012.04.005

Li, G. C., Huang, G. H., Lin, Q. G., Cai, Y. P., Chen, Y. M., & Zhang, X. D. (2012). Development of an interval multi-stage stochastic programming model for regional energy systems planning and GHG emission control under uncertainty. *International Journal of Energy Research*, 36(12), 1161-1174. doi: 10.1002/er.1867

Li, M. W., Li, Y. P., & Huang, G. H. (2011). An interval-fuzzy two-stage stochastic programming model for planning carbon dioxide trading under uncertainty. *Energy*, 36(9), 5677-5689. doi: 10.1016/j.energy.2011.06.058

Li, M., Guo, P., Fang, S. Q., & Zhang, L. D. (2013). An inexact fuzzy parameter two-stage stochastic programming model for irrigation water allocation under uncertainty. *Stochastic Environmental Research and Risk Assessment*, 27(6), 1441-1452. doi: 10.1007/s00477-012-0681-y

Li, Y. F., Li, Y. P., Huang, G. H., & Chen, X. (2010). Energy and environmental systems planning under uncertainty-An inexact fuzzy-stochastic programming approach. *Applied Energy*, 87(10), 3189-3211. doi: 10.1016/j.apenergy.2010.02.030

Li, Y. P., & Huang, G. H. (2010). Inexact joint-probabilistic stochastic programming for water resources management under uncertainty. *Engineering Optimization*, 42(11), 1023-1037. doi:

10.1080/03052151003622539

Liu, P., Pistikopoulos, E. N., & Li, Z. (2010). Decomposition Based Stochastic Programming Approach for Polygeneration Energy Systems Design under Uncertainty. *Industrial & Engineering Chemistry Research*, 49(7), 3295-3305. doi: 10.1021/ie901490g

Ma, Z., Kwon, R. H., & Lee, C. G. (2010). A stochastic programming winner determination model for truckload procurement under shipment uncertainty. *Transportation Research Part E-Logistics and Transportation Review*, 46(1), 49-60. doi: 10.1016/j.tre.2009.02.002

Mohammadi, S., Mozafari, B., & Soleymani, S. (2013). A stochastic programming approach for optimal microgrid economic operation under uncertainty using 2m+1 point estimate method. *Journal of Renewable and Sustainable Energy*, 5(3). doi: 10.1063/1.4808039

Mousavi, S. M., Vandani, B., Tavakkoli-Moghaddam, R., & Hashemi, H. (2014). Location of cross-docking centers and vehicle routing scheduling under uncertainty: A fuzzy possibilistic-stochastic programming model. *Applied Mathematical Modelling*, 38(7-8), 2249-2264. doi: 10.1016/j.apm.2013.10.029

Nocedal, J. (1999) Numerical Optimization.Secaucus, NJ, USA: Springer, p 23.

Nocedal, J., Waechter, A., Waltz, R.A. (2005) Adaptive barrier strategies for nonlinear interior methods. Technical Report RC 23563, IBM T.J. Watson Research Center, Yorktown Heights, USA, March 2005.

Palmer, A. (2004) The environmental implications of grocery home delivery, ELA doctorate workshop, Centre for Logistics and Supply Chain Management, Cranfield University

Pardalos, P.M. (Ed.). (2002) Combinatorial and Global Optimization. River Edge, NJ, USA: World Scientific. p 10pq.

Pardalos, P.M. (Ed.). (2004) Supply Chain and Finance. Singapore: World Scientific Publishing Company, Incorporated. p v.

Prékopa, A. (1995) Stochastic programming. Dordrecht, The Netherlands: Kluwer Academic Publishers.

Pronello, C., André, M. (2000) Pollutant emissions estimation in road transport models. INRETS-LTE report, vol. 2007

Sahinidis, N.V. (2004) Optimization under uncertainty: state-of-the-art and opportunities, *Computers & Chemical Engineering*, 28(6-7): 971-983. doi: 10.1016/j.compchemeng.2003.09.017

Sarker, R. (Ed.). (2002) Evolutionary Optimization.Secaucus, NJ, USA: Kluwer Academic Publishers, p 4.

Sbihi, A., Eglese, R.W. (2007) Combinatorial optimization and Green logistics. *4OR: A Quarterly Journal of Operations Research*, 5(2), pp. 99-116.

Simchi-Levi, D. (2004) Managing the Supply Chain. Blacklick, OH, USA: McGraw-Hill Professional Publishing. p 2-3.

Singh, A. R., Jain, R., & Mishra, P. K. (2013). Capacities-based supply chain network design considering demand uncertainty using two-stage stochastic programming. *International Journal of Advanced Manufacturing Technology*, 69(1-4), 555-562. doi: 10.1007/s00170-013-5054-2

Sodhi, M. S., & Tang, C. S. (2011). Determining supply requirement in the sales-and-operations-planning (S&OP) process under demand uncertainty: a stochastic programming formulation and a spreadsheet implementation. *Journal of the Operational Research Society*, 62(3), 526-536. doi: 10.1057/jors.2010.93

Tan, Q. A., Huang, G. H., & Cai, Y. P. (2010). A Superiority-Inferiority-Based Inexact Fuzzy Stochastic Programming Approach for Solid Waste Management Under Uncertainty. *Environmental Modeling & Assessment*, 15(5), 381-396. doi: 10.1007/s10666-009-9214-6

Tintner, G. (1955) Stochastic linear programming with applications to agricultural economics. In H. A. Antosiewicz (Ed.), Proceedings of the Second Symposium in Linear Programming (pp. 197-228), National Bureau of Standards, Washington, DC.

Tits, A. L., Wächter, A., Bakhtiari, S., Urban, T. J., Lawrence, C. T. (2003) A primal-dual interior-point method for nonlinear programming with strong global and local convergence properties. *SIAM Journal on Optimization*, 14 (1), 173-199

Waechter A., Biegler, L.T. (2005) Line search filter methods for nonlinear programming: Local convergence. *SIAM Journal on Optimization*, 16(1):32-48

Waechter A., Biegler, L.T. (2006) On the implementation of a primal-dual interior point filter line search algorithm for large-scale nonlinear programming. *Mathematical Programming*, 106(1):25-57

Waechter, A. (2002) An Interior Point Algorithm for Large-Scale Nonlinear Optimization with Applications in Process Engineering. Ph.D. thesis, Carnegie Mellon University, Pittsburgh, PA, USA, January 2002.

Waechter, A., Biegler, L.T. (2005) Line search filter methods for nonlinear programming: Motivation and global convergence. *SIAM Journal on Optimization*, 16(1):1-31

Xie, Y. L., Li, Y. P., Huang, G. H., & Li, Y. F. (2010). An interval fixed-mix stochastic programming method for greenhouse gas mitigation in energy systems under uncertainty. *Energy*, 35(12), 4627-4644. doi: 10.1016/j.energy.2010.09.045

Xu, J. P., & Wu, D. D. (2010). Sustainable development and planning of coal industry under uncertainty using system dynamic and stochastic programming. *International Journal of Environment and Pollution*, 42(4), 371-387.

Yamashita, H. (1998) A globally convergent primal-dual interior-point method for constrained optimization. *Optimization Methods and Software*, 10, 443–469

Zanjani, M. K., Nourelfath, M., & Ait-Kadi, D. (2010). A multi-stage stochastic programming approach for production planning with uncertainty in the quality of raw materials and demand. *International Journal of Production Research*, 48(16), 4701-4723. doi: 10.1080/00207540903055727

Zimmermann, H.-J. (1991) Fuzzy set theory and its application (2nd ed.). Boston: Kluwer Academic Publishers.

Chapter 6. Transportation Network Design by Heuristic Methods

Abstract. This study suggests and employs an optimization method; the Genetic Algorithm, on the problems of the selected transportation systems, by taking into consideration the substantial environmental effects of these transportation systems. Several environmental impacts and logistic system costs can be reduced by applying optimal solutions and putting into application the heuristic methods. Computational solutions disclose that applied methods are effective, flexible, and easy to enforce in solving problems. This thence reduce ecological effects on particular logistical actions resulting from inefficient network designs. This study establishes that heuristics methods have the ability to provide optimal solutions for the problems associated with logistic systems with the aim of reducing environmental impacts on particular logistical actions with ineffective network designs.

Keywords: genetic algorithm, transportation costs, environmental effects

1. Introduction

Transport sector is one of the few sectors, which have ever-increasing greenhouse gas (GHG) emissions. According to IPCC reports, the transport sector accounts for 13.1% of GHG emissions, 23% of global carbon dioxide (CO_2) emissions, and 26% of total world energy use. The transport sector is one of the constantly growing sectors.

The growth of the transport sector heavily depends on fossil fuels and imports. Thus, with the growing of transport sector, the demand on oil is expected to increase in the years to come. Consumption data of fuel is one of the reliable elements used to calculate emission levels.

On the other hand, from the global scale based on the estimates of the years between 2000 and 2050, the International Energy Agency (IEA) warns that there is a possibility of 50% increase in CO_2 emissions originating from transport sector.

The next section of this paper covers the general literature. The third section looks at the fundamental combinatorial optimization background as well as details of genetic algorithm (GA) method and fundamental routing problems are exhibited with designated constraints. In the same section, scenario based travelling salesmen (TSP) path problems with randomly assigned locations are used and solved by the heuristic methods and a fundamental reason for choosing the GA method for problems are briefly given. Finally, this paper is summarized in section 4.

2. Literature review

Practices correlated with environmental issues and performances constitute both the internal and external activities (Cousins 2006). For instance, the more products are stored or shipped in a given cubic capacity, the more the associated unit costs. This implies that the environmental impact may be cut down. Packaging in the form of condensed containers, such as pallets, or recyclable containers, will often necessitate returning them to the point of origin for them to be reused.

For logisticians, the problem manifests itself in the form of reverse logistics. Waste packaging requires to be taken back up the supply chain (Rushton 2006). According to Stock (2002), to reduce environmental impacts, some typical measures in logistics activities are, but not limited to:

- miles per gallon/liters per kilometer of fuel used,
- per centum of fleet using less polluting fuels,
- utilization of vehicle load space expressed as a percentage,
- per centum of empty miles or kilometers run by vehicles,
- targets for reducing waste packaging,
- targets for reducing noise levels

Kutz (2003) states that in the design process, the main emphasis lies on defining a network that connects and/or links up selected nodes with a certain quality. It is taken for granted that the accessibility of these nodes is thereby assured. Of adequate importance is the quality of the designed network as well as a corroboration of other sustainability criteria, such as:

- The environment and livability,
- Traffic safety,
- Network accessibility,
- Economic accessibility, and
- The Costs.

Environment and livability effects are emissions of harmful substances to man, fauna, and flora; noise nuisance; and fragmenting the landscape. Traffic emits a number of harmful substances such as CO, CO_2, C_xH_y, NO_x, Pb, and particulate matter. The size of these emissions depends on various factors, such as fuel usage, type of fuel, speed, driving cycle, and gradient. One has to employ a highly complex method of calculation in order to take account of all these factors (Kutz 2003). Critical cognizance and responsiveness to the situation is crucial.

However, environmental effects tend to be deemed secondary objectives besides the primary criteria of the cost, quality, and delivery at supply chains and logistics units of some organizations. Pressure to consider an environmental prospect from legislation is not the only business concern (Stock 2002). A "green" image has become an important marketing opportunity. Thus, it is imperative for businesses to revisit their supply chain performance metrics in response to developing external institutional pressure as well as the clients' environmental requirements (Stock 2002).

Grounded on the sort of decision variables, the most noteworthy network design problems can be separated into distinct and continuous models according to Gallo *et al.* (2010) and Beltran *et al.* (2009).

- Continuous variable models were developed and formulated in the papers by Dantzig *et al.* (1979), Abdulaal and Le Blanc (1979), Marcotte (1983), Harker and Friesz (1984), Le Blanc and Boyce (1986), Suwansirikul *et al.* (1987), Friesz *et al.* (1992), Davis (1994), Cho and Lo (1999), Meng *et al.* (2001), Meng and Yang (2002), and Chiou (2005).
- Discrete variable models were developed and formulated in the papers by Billheimer and Gray (1973), Le Blanc (1975), Los (1979), Boyce and Janson (1980), Foulds (1981), Los and Lardinois (1982), Poorzahedy and Turnquist (1982), Chen and Alfa (1991), Herrmann *et al.* (1996), Solanki *et al.* (1998), Cruz *et al.* (1999), Drezner and Wesolowsky (2003), Gao *et al.* (2005), Poorzahedy and Abulghasemi (2005), Poorzahedy and Rouhani (2007) and Ukkusuri *et al.* (2007).

For network design problems, poly-criteria proficiency for urban networks with the usage of genetic algorithm proposed by Pattnaik *et al.* (1998), Dhingra *et al.* (2000), Ngamchai and Lovell (2003), Cantarella and Vitetta (2006), Russo and Vitetta (2006). Cantarella (2006) also suggests other methods, such as Simulated Annealing (SA), Tabu Search (TS), Path Relinking, Climbing, and Genetic Algorithms. On account of the non-convexity of the transportation network design problem as reported by Newell (1979), the most beneficial and effective solution methods are based on heuristic procedures. For the other most noteworthy works about network design, it is possible to mention Baaj and Mahmassani (1992, 1995), Ceder and Israeli (1993) and Carrese and Gori (2002).

Table 1. Some highlighted literature about transportation network design

A bilevel flow model for hazmat transportation network design (Bianco et al. 2009).
Model for Microcirculation Transportation Network Design (Chen and Shi 2012).
A review of urban transportation network design problems (Farahani et al. 2013).
Multimodal Freight Transportation Network Design Problem for Reduction of Greenhouse Gas Emissions (Kim et al. 2013).
Global optimization method for mixed transportation network design problem: A mixed-integer linear programming approach (Luathep et al. 2011).
System-optimal stochastic transportation network design (Patil and Ukkusuri 2007).
Mixed Transportation Network Design under a Sustainable Development Perspective (Qin et al. 2013).
Approximation Techniques for Transportation Network Design Problem under Demand Uncertainty (Sharma et al. 2011).
Pareto Optimal Multiobjective Optimization for Robust Transportation Network Design Problem (Sharma et al. 2009).
Robust transportation network design under demand uncertainty (Ukkusuri et al. 2007).
Multi-period transportation network design under demand uncertainty (Ukkusuri and Patil 2009).
Freight Transportation Network Design Problem for Maximizing Throughput under Uncertainty (Unnikrishnan and Waller 2009).
Transportation Network Design considering Morning and Evening Peak-Hour Demands (Wang et al. 2014).
Hazmats Transportation Network Design Model with Emergency Response under Complex Fuzzy Environment (Xu et al. 2013).
Bilevel programming model and solution method for mixed transportation network design problem (Zhang and Gao 2009).

3. Data and methods

A combinatorial optimization problem can be written as

$$x^* = \min_{x \in D \subseteq X} f(x), \quad (1)$$

where the objective is to find $x^* \in D \subseteq X$. X is bounded by a finite space and $D \subseteq X$ is the subspace of feasible solutions. $f : X \to R^1$ is the objective function.

To obtain solutions for the types of problems, as shown in (1), there are several approaches.

Routing problem I: The problem is to ascertain the operation plan satisfying the demand at various zones at minimum cost.

Objective function is

$$\text{Minimize } f_{obj} = \sum_{i=1}^{G} \sum_{j=1}^{Z} \sum_{k=1}^{F} C_{ijk} x_{ijk}$$

Subject to

$$\sum_{i=1}^{G} \sum_{k=1}^{F} L_k x_{ijk} \geq D_j \qquad \forall j$$

$$\sum_{j=1}^{Z} \sum_{k=1}^{F} L_k x_{ijk} \leq S_j \qquad \forall i$$

$$\sum_{j=1}^{Z} L_k x_{ijk} \leq U_{ki} \qquad \forall k, i$$

$$x_{ijk} \geq 0 \qquad \forall i, j, k$$

where parameters are

G = Number of source locations (index i)

Z = Number of receiving nodes for containers (index j)

F = Number of trailers available (index k)

L_k = Load capacity of trailer k

S_i = Quantity of available containers for transportation from location i

D_j = Quantity of containers required by zone j

C_{ijk} = Unit cost of transporting from location i to zone j by trailer k

U_{ik} = Maximum allowable containers that can be transported from location i by trailer k in a given period

and variables

x_{ijk} = the number of trips required by trailer k from location i to zone j

Routing Problem II: A generic model that practitioners encounter in many planning and decision processes. For instance, the delivery and collection of containers/cargos, etc.

Objective function is

$$\text{Minimize } Z = \sum_{k=1}^{K} \sum_{(i,j) \in A} C_{ij} x_{kij}$$

Subject to

$$\sum_{i=1}^{n} y_{ij} = 1, \quad j = 2, 3, \ldots, n$$

$$\sum_{j=1}^{n} y_{ij} = 1, \quad i = 2, 3, \ldots, n$$

$$\sum_{j=1}^{n} y_{1j} = K$$

$$\sum_{j=1}^{n} y_{i1} = K$$

$$\sum_{i=1}^{n}\sum_{j=2}^{n} D_j x_{kij} \leq U, \quad k = 1, 2, \ldots, K$$

$$\sum_{k=1}^{K} x_{kij} = y_{ij} \quad \forall i, j$$

$$\sum_{(i,j)\in S \times S} y_{ij} \leq |S|-1, \quad \text{for all subsets } S \text{ of } \{2,3,\ldots,n\}$$

$$x_{kij} = 0 \text{ or } 1 \quad \forall (i,j) \in A \text{ and } \forall k$$

$$y_{ij} = 0 \text{ or } 1 \quad \forall (i,j) \in A$$

- A fleet of M capacitated vehicles located in a depot (i=1)
- A set of target zones (of size N-1), each having a demand Dj (j=2,...,N)
- A cost Cij of traveling from location i to location j
- The problem is to find a set of routes for delivering / picking up goods to/from the target zones at minimum possible cost.

The vehicle fleet is homogeneous and that each vehicle has a capacity of U units.

and variables:

x_{kij} = 1 if the vehicle k travels on the arc i to j, 0 otherwise

y_{ij} = 1 if any vehicle travels on the arc (i,j), 0 otherwise

3.1 Various scenarios and network configurations – Optimal solutions by heuristics methods

In this section, several potential scenarios are considered while designing a transport network. On arbitrarily distributed locations, how heuristics methods can be employed to meet particular transport network design needs are considered. Developed scenarios are based on a theoretical assumption that city locations are all accessible from every direction without geographical limitation, other constraints, and route affecting factors. Therefore, founded on the supposition, five scenarios are introduced. Scenarios 1 to 4 are formulated using genetic algorithm (See Figure 1). GA network designs are generated and simulated by utilizing Matlab® codes.

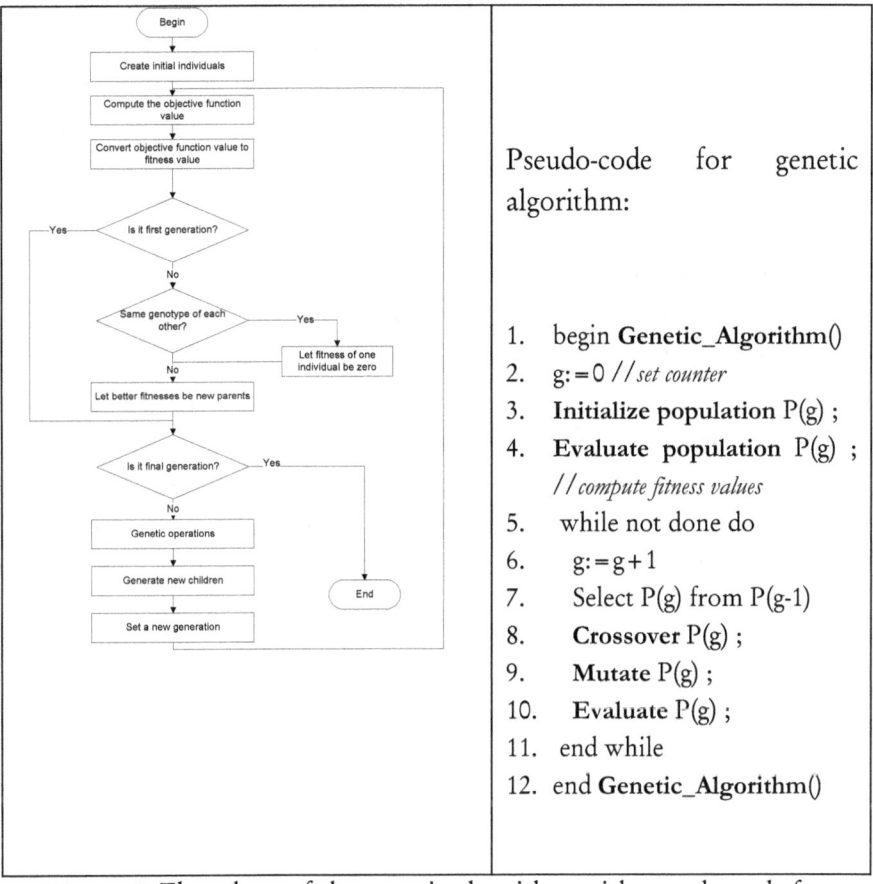

Pseudo-code for genetic algorithm:

1. begin **Genetic_Algorithm()**
2. g:=0 // *set counter*
3. **Initialize** population P(g) ;
4. **Evaluate** population P(g) ;
   // *compute fitness values*
5. while not done do
6. g:=g+1
7. **Select** P(g) from P(g-1)
8. **Crossover** P(g) ;
9. **Mutate** P(g) ;
10. **Evaluate** P(g) ;
11. end while
12. end **Genetic_Algorithm()**

Figure 1. Flowchart of the genetic algorithm with pseudo-code for genetic algorithm

Scenario 1

In this scenario, closed loop transportation networks for 60 city locations with three trucks are taken into consideration. An arbitrarily distributed city locations on an XY plane and fixed number of vehicles, e.g. 3 trucks, are given a task to visit all of their assigned locations as such that the total distance travelled by these vehicles will be minimum.

To downplay the total distance, GA method is applied to find out the assigned locations for each vehicle and to keep the travelled distance for these vehicles at minimum. Based on the algorithm's outcome, minimum total distance for three vehicles can only be achieved by the network configuration shown at Figure 6a. The solution network design (Figure 2a) is based on the randomly selected and predetermined city locations (see top left of the Figure 2b). The best solution history (bottom right of Figure 2b) is the gradually reduction of the total distance at each iteration.

Optimization of Logistics: Theory and Practice

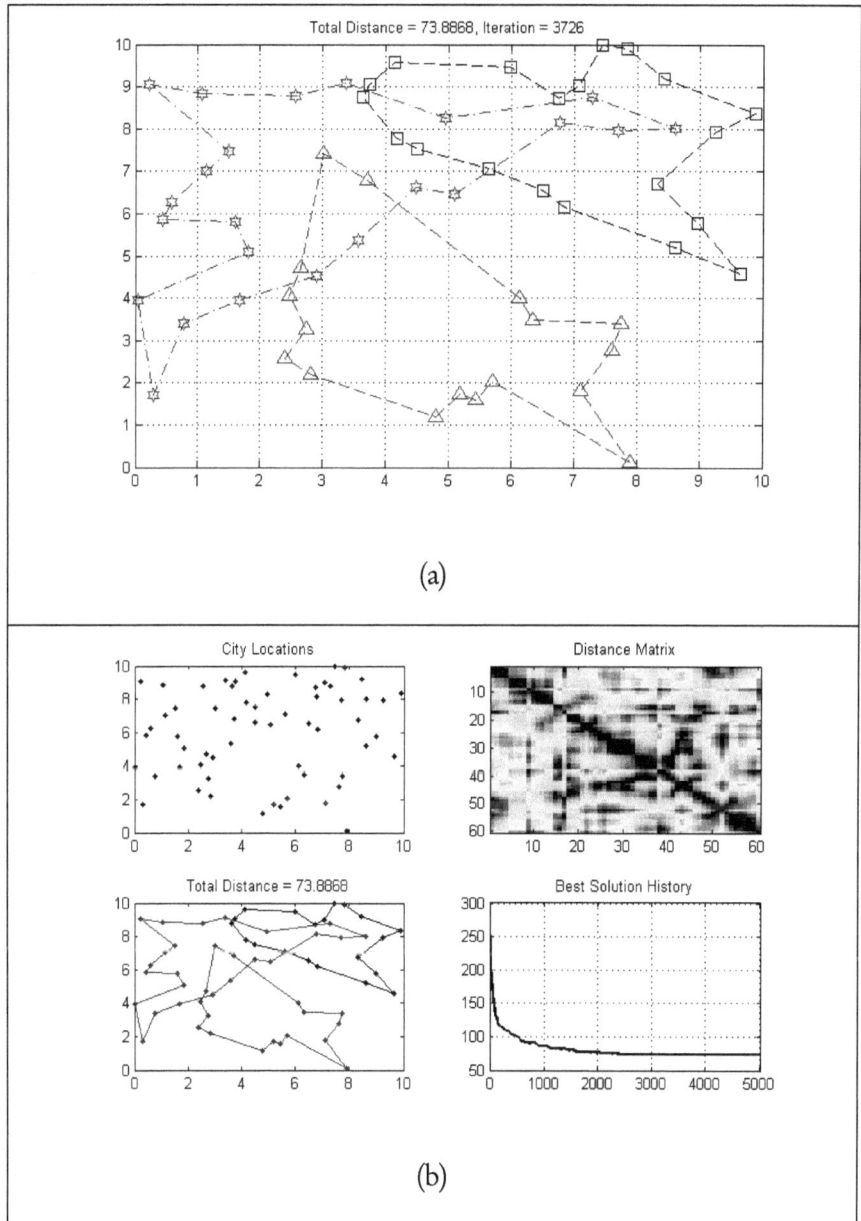

Figure 2. Three trucks travel at minimum possible total distance. All locations visited.

Scenario 2

In this scenario, a transportation network for 60 city locations with three trucks are considered with a fixed start node. On a randomly distributed city locations, vehicles are assigned to specific city locations by the GA algorithm to minimize the total travelling distance of the vehicles.

A closed loop network for each vehicle has been generated. The total distance can only be minimized by the configuration shown at Figure 3b.At each iteration of the algorithm, the total distance reduces to a global minimum as shown at the best solution history (see bottom right of the Figure 3b).

Optimization of Logistics: Theory and Practice

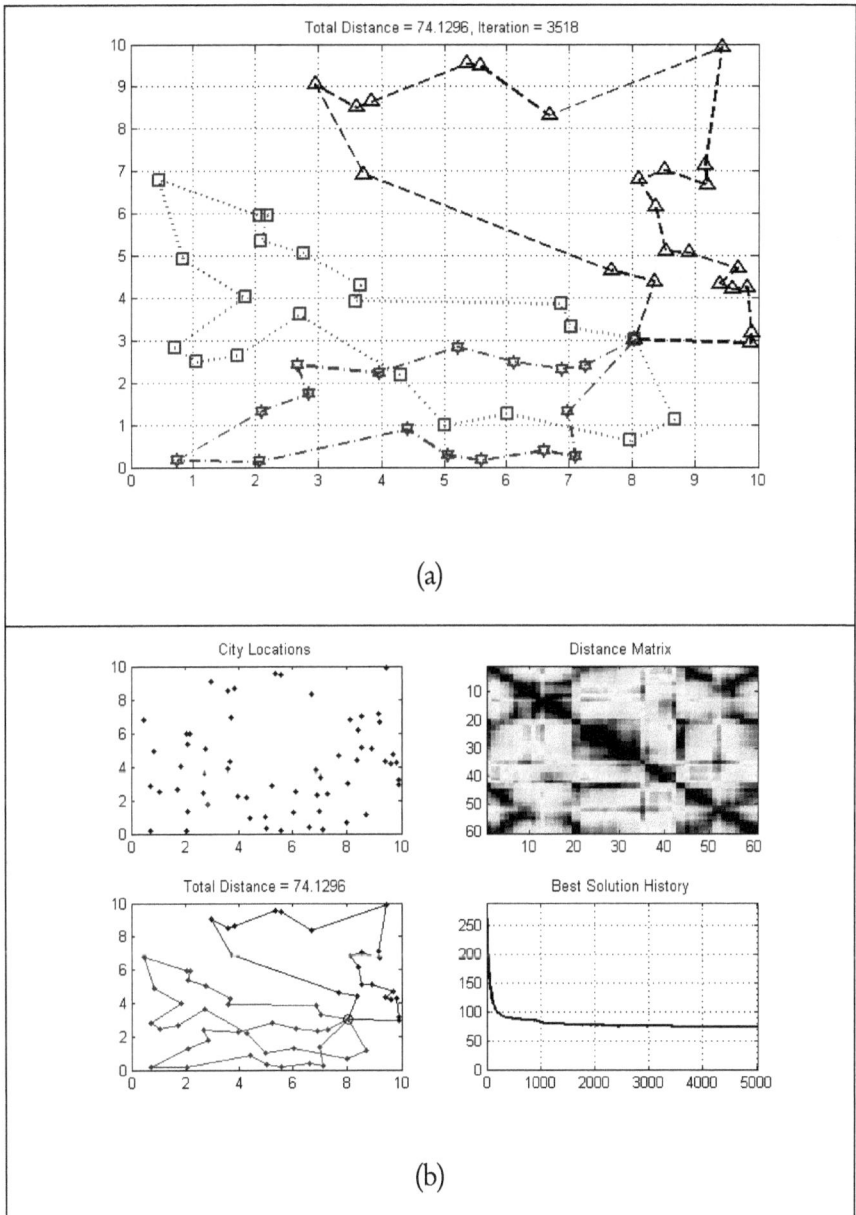

Figure 3. Three trucks, starting from a fixed location, travel at minimum possible total distance. All locations visited.

Scenario 3

A scenario of a fixed start point with open ends for three trucks of 60 randomly distributed city locations is considered. All locations are covered on the transportation network. Minimum total distance can only be achieved by the network configuration shown at the Figure 4a.

"Fixed start-open ends" configuration shown at the Figure 4a provides the minimum possible total distance for all the vehicles. Other details of the configuration can be seen at the Figure 4b.

Optimization of Logistics: Theory and Practice

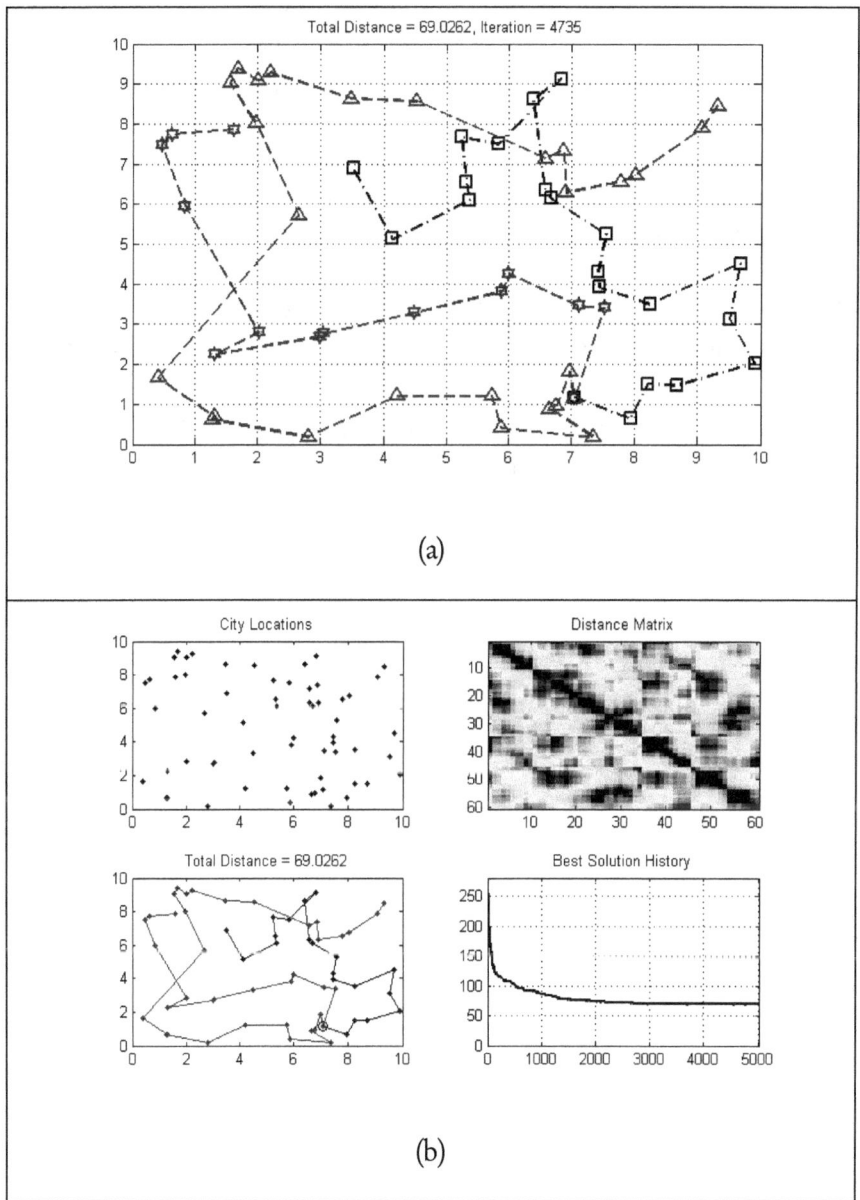

Figure 4. Three trucks, starting from a fixed location, travel at minimum possible total distance. All locations visited.

Scenario 4

This scenario considers the basic minimum total distance for *fixed start* and *fixed end* location for a travelling vehicle. Minimum total distance by visiting all the locations can only be achieved by following the route shown at the Figure 5a.

All locations are visited with a minimum total distance. Other details of the configuration can be seen at the Figure 5b.

In addition, based on all the above assumptions, comparing the algorithms for network design performance from different theoretical and empirical categories is a complicated task. Owing to the fact that there is not a specific empirical baseline, that enables unbiased comparison among algorithms.

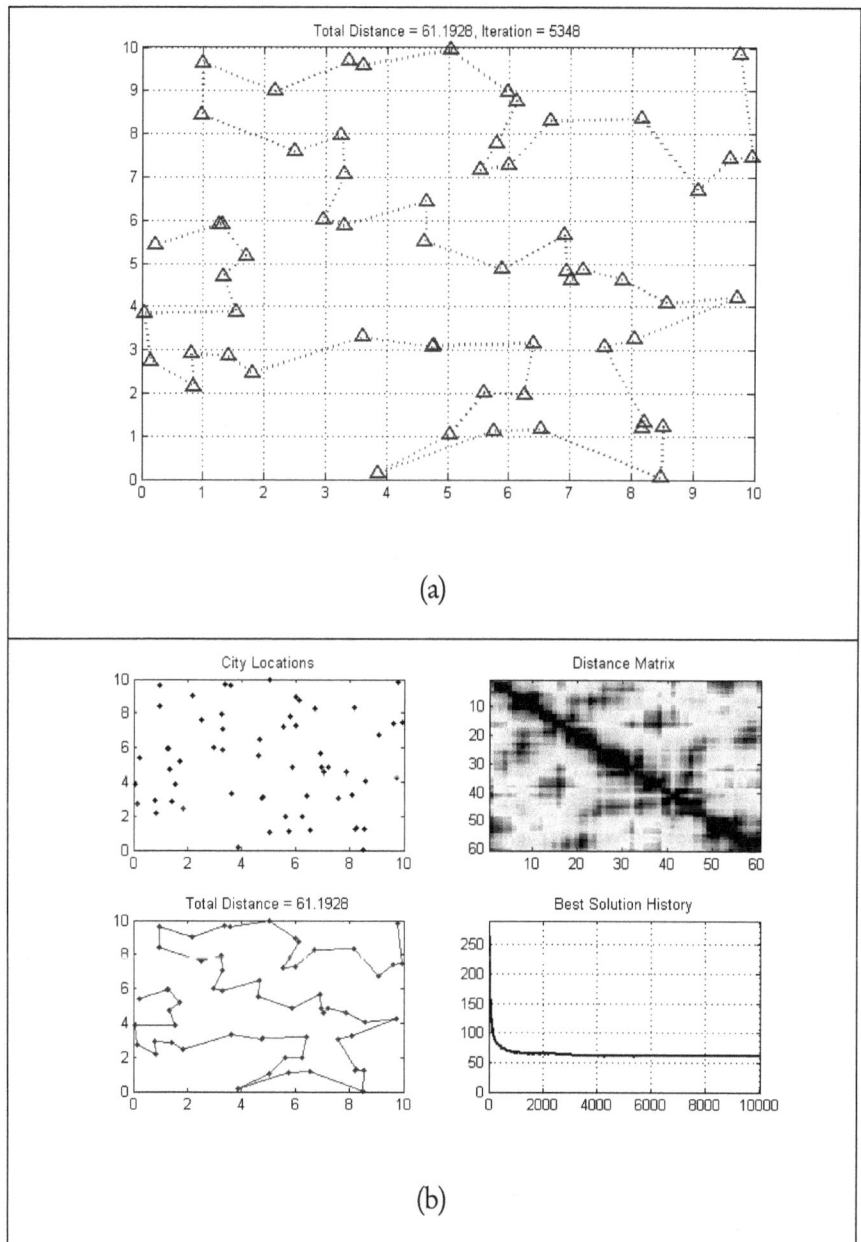

Figure 5. Minimum possible total distance by covering all locations on the route. 1 vehicle with a fixed start and fixed end locations.

4. Research findings and discussions

Additional meta-heuristic algorithms such as Simulated Annealing (SA), Tabu Search (TS), Ant Colony Optimization (ACO), Particle Swarm Optimization (PSO), Memetic Algorithms (MA), etc. are rather mutual in solving several forms of problems. On the other hand, there are important differences among them. These differences originate from theoretical and empirical grounds of algorithms (Aarts and Korst 1989; Goldberg 1989; Dorigo *et al.* 1999; Ehrgott 2002).

Measuring and quantifying environmental pollutants based on the design of network has not been stressed in literature. It is therefore a difficult and even practically inconceivable task to achieve. In a report by Pronello and Andrè (2000), it is indicated that pollution induced by a set of vehicle routes can be treacherous. However, it is unmistakable that shortening the total distance and thence utilizing optimal routes by heuristics algorithms will provide environmental benefits due to the reduction in fuel consumption and the consequent pollutants (Sbihi and Eglese 2007).

5. Conclusion

While there is no single efficacious solution that conforms to all transport emission reduction problems, the optimization method presented on this paper is satisfactorily proposed as route optimization methods that are theoretically applied. By applying optimal solutions and putting into practice the heuristics algorithm method such as Genetic Algorithm, numerous environmental impacts as well as an entire logistics system cost can be cut down. Computational results disclose that the Applied Method is efficient, versatile, and easy to use in solving problems and reducing ecological effects of particular logistical actions with inefficient network designs.

Generally, based on the number and complexity of transportation networks, obtaining optimal solutions with heuristic methods is a non-deterministic polynomial-time (NP) hard problem and computational time exponentially increases contingent to the number of resources involved in the problem. It is stated in particular with detail on this paper that heuristics algorithm's approach gives stable solutions by finding out optimal values. By utilizing and running the proposed high performing algorithms for the problems of transportation network designs, it is apparent that there will be significant reduction in emissions along with the total operational costs.

Lastly, it has been demonstrated that heuristics methods are capable of providing optimal solutions to the problems of logistics systems with the concerns of reducing environmental influences of particular logistical actions with inefficient network designs. As green issues in logistical and supply-chain systems have been receiving a growing concentration from the last decade, the proposed solutions by using heuristic methods to the environmental issues will find encouraging areas on network design problems.

6. References

Aarts, E., Korst, J., (1989) Simulated Annealing and Boltzmann Machines. Wiley. 284 p.

Abdulaal, M., Le Blanc, L., (1979) Continuous equilibrium network design models, *Transportation Research Part B*, 13 (1), 19-32.

Baaj, H., Mahmassani, H.S., (1992) TRUST: a lisp program for the analysis of transit route configurations, *Transportation Research Record*, 1283, 125-135.

Baaj, H., Mahmassani, H.S., (1995) Hybrid route generation heuristic algorithm for the design of transit networks, *Transportation Research Part C*, 3 (1), 31-50.

Beltran, B., Carrese, S., Cipriani, E., Petrelli, M. (2009) Transit network design with allocation of green vehicles: A genetic algorithm approach, *Transportation Research Part C: Emerging Technologies*, 17 (5), 475-483.

Bianco, L., Caramia, M., & Giordani, S. (2009). A bilevel flow model for hazmat transportation network design. *Transportation Research Part C-Emerging Technologies*, 17(2), 175-196. doi: 10.1016/j.trc.2008.10.001

Billheimer, J.W., Gray, P., (1973) Network design with fixed and variable cost elements, *Transportation Science*, 7 (1), 49-74.

Boyce, D.E., Janson, B.N., (1980) A discrete transportation network design problem with combined trip distribution and assignment,

Transportation Research Part B, 14 (1-2), 147–154.

Cantarella, G.E., Pavone, G., Vitetta, A., (2006) Heuristics for urban road network design: Lane layout and signal settings, *European Journal of Operational Research*, 175 (3), 1682–1695.

Cantarella, G.E., Vitetta, A., (2006) The multi-criteria road network design problem in an urban area, *Transportation*, 33 (6), 567–588.

Carrese, S., Gori, S., (2002) An urban bus network design procedure. In: Patriksson, M., Labbè, M., (Eds.), Transportation Planning: State of the Art, 177–196.

Ceder, A., Israeli, Y., (1993) Design and evaluation of transit routes in urban networks. In: Proceedings of the 3rd International Conference on Competition and Ownership in Surface Passenger Transport, Ontario, Canada.

Chen, M. and Alfa, A.S., (1991) A network design algorithm using a stochastic incremental traffic assignment approach, *Transportation Science*, 25 (3), 215–224.

Chen, Q., & Shi, F. (2012). Model for Microcirculation Transportation Network Design. *Mathematical Problems in Engineering*. doi: 10.1155/2012/379867

Chiou, S.W., (2005) Bilevel programming for the continuous transport network design problem, *Transportation Research Part B*, 39 (4), 361–383.

Cho, H.J., Lo, S.C., (1999) Solving bilevel network design problem using linear reaction function without nondegeneracy assumption, *Transportation Research Record*, 1667, 96–106.

Claudio B. Cunha, Silva M.R., (2007) A genetic algorithm for the problem of configuring a hub-and-spoke network for a LTL trucking company in Brazil, *European Journal of Operational Research*, Volume 179, Issue 3, 16, Pages 747-758

Colorni, A., Dorigo, M., Maffioli, F., Maniezzo, V., Righini, G., Trubian, M., (1996) Heuristics from nature for hard combinatorial problems. *International Transactions in Operational Research*, 3 (1), 1–21.

Cousins, Paul D. (Ed.), (2006) Supply Chain Management Theory and Practice : The Emergence of an Academic Discipline? Bradford, GBR: Emerald Group Publishing Limited, p 797.

Cruz, F.R.B., Smith, J.M.G., Mateus, G.R., (1999) Algorithms for a multi-level network optimization problem, *European Journal of Operational Research*, 118 (1), 164–180.

Dantzig, G.B., Harvey, R.P., Lansdowne, Z.F., Robinson, D.W., Maier, S.F., (1979) Formulating and solving the network design problem by decomposition, *Transportation Research Part B*, 13 (1), 5-17.

Davis, G.A., (1994) Exact local solution of the continuous network design problem via stochastic user equilibrium assignment, *Transportation Research Part B*, 28 (1), 61-75.

Dhingra, S.L., Muralidhar S. & Krishna Rao K.V., (2000) Public transport routing and scheduling using genetic algorithms. In: Proceedings Presented at the CASPT 8th International Conference, Berlin, Germany.

Dorigo, M., Di Caro, G., Gambardella, L. M. (1999) Ant algorithms for discrete optimization. *Artificial life*, 5 (2): 137-172.

Drezner, Z., Wesolowsky, G.O., (2003) Network design: Selection and design of links and facility location, *Transportation Research Part A*, 37 (3), 241-256.

Ehrgott, M., Gandibleux, X. (eds.), (2002) Multiple Criteria Optimization: State of the Art Annotated Bibliographic Surveys. Secaucus, NJ, USA: Kluwer Academic Publishers. 1st edition. Springer. 520 p.

Farahani, R. Z., Miandoabchi, E., Szeto, W. Y., & Rashidi, H. (2013). A review of urban transportation network design problems. *European Journal of Operational Research*, 229(2), 281-302. doi: 10.1016/j.ejor.2013.01.001

Foulds, R.L., (1981) A multicommodity flow network design problem, *Transportation Research Part B*, 15 (4), 273-283.

Friesz, T.L., Cho, Hsun-Jung, Mehta, Nihal J., Tobin, Roger L., Anandalingam, G., (1992) A simulated annealing approach to the network design problem with variational inequality constraints, *Transportation Science*, 26 (1), 18-26.

Gallo, M., D'Acierno, L., Montella, B., (2010) A meta-heuristic approach for solving the Urban Network Design Problem, *European Journal of Operational Research*, 201 (1), 144-157

Gao, Z., Wu, J., Sun, H., (2005) Solution algorithm for the bi-level discrete network design problem, *Transportation Research Part B*, 39 (6), 479-495.

Garey and Johnson, (1979) Computer and Intractability,Freeman, San Fransico, CA.

Glover, Fred (Ed.), (2002) Handbook of Metaheuristics. Secaucus, NJ, USA: Kluwer Academic Publishers, p 55.

Goldberg, D., (1989) Genetic Algorithms in Search, Optimization and Machine Learning. Addison Wesley.

Harker, T.P., Friesz, T.L., (1984) Bounding the solution of the continuous equilibrium network design problem. In: Volmuller, J., Hamerslag, R. (Eds.), Proceedings of 9th International Symposium on Transportation and Traffic Theory. VNU Science Press, Utrecht, The Netherlands, 233–252.

Herrmann, J.W., Ioannou, G., Minis, I., Proth, J.M. (1996) A dual ascent approach to the fixed-charge capacitated network design problem, *European Journal of Operational Research*, 95 (3), 476–490.

IEA, (2001) Saving Oil and Reducing CO2 Emissions in Transport. International Energy Agency, OECD, 194 pp.

IEA, (2002a) Transportation Projections in OECD Regions – Detailed report. International Energy Agency, 164 pp.

IEA, (2002b) Bus Systems for the Future: Achieving Sustainable Transport Worldwide. International Energy Agency, 188 pp.

IEA, (2003) Transport Technologies and policies for energy security and CO2 Reductions. Energy technology policy and collaboration papers, International Energy Agency, ETPC paper no 02/2003.

IEA, (2004a) World Energy Outlook 2004. International Energy Agency, 570 pp.

IEA, (2004b) Energy Technologies for a Sustainable Future: Transport. International Energy Agency, Technology Brief, 40 pp.

IEA, (2004c) Biofuels for Transport: An International Perspective. International Energy Agency, Paris, 210 pp.

IEA, (2004d) Reducing Oil Consumption in Transport - Combining Three Approaches.

IEA, (2004d) Reducing Oil Consumption in Transport - Combining Three Approaches. IEA/EET working paper by L. Fulton, International Energy Agency, Paris, 24 pp. Jones, M. Tim, 2003. AI Application Programming. Herndon, VA, USA: Charles River Media, p 115.

IEA, (2005) Prospects for Hydrogen and Fuel Cells. International Energy Agency, Paris, 253 pp.

IEA, (2006a) Energy Technology Perspectives 2006; Scenarios & Strategies to 2050. International Energy Agency, Paris, 479 pp. 383

Kim, S., Park, M., & Lee, C. (2013). Multimodal Freight Transportation Network Design Problem for Reduction of Greenhouse Gas Emissions. *Transportation Research Record*, (2340), 74-83. doi: 10.3141/2340-09

Kutz, Myer. (2003) Handbook of Transportation Engineering. New York, NY, USA: McGraw-Hill. p 69-70.

Le Blanc, L.J., (1975) An algorithm for the discrete network design problem, *Transportation Science*, 9 (3), 183–199.

Le Blanc, L.J., Boyce, D.E., (1986) A bilevel programming algorithm for exact solution of the network design problem with user optimal flows, *Transportation Research Part B*, 20, 259–265.

Los, M. and Lardinois, C., (1982) Combinatorial programming, statistical optimization and the optimal transportation network problem, *Transportation Research Part B*, 16 (2), 89–124.

Los, M., (1979) A discrete-convex programming approach to the simultaneous optimization of land use and transportation, *Transportation Research Part B*, 13 (1), 33–48.

Luathep, P., Sumalee, A., Lam, W. H. K., Li, Z. C., & Lo, H. K. (2011). Global optimization method for mixed transportation network design problem: A mixed-integer linear programming approach. *Transportation Research Part B-Methodological*, 45(5), 808-827. doi: 10.1016/j.trb.2011.02.002

Meng, Q, Yang, H., (2002) Benefit distribution and equity in road network design, *Transportation Research Part B*, 36, 19–35.

Meng, Q., Yang, H., Bell, M.G.H., (2001) An equivalent continuously differentiable model and a locally convergent algorithm for the continuous network design problem, *Transportation Research Part B*, 35 (1), 83–105.

Newell, G.F., (1979) Some issue relating to the optimal design of bus lines, *Transportation Science*, 13 (1), 20–35.

Ngamchai, S., Lovell, D.J., (2003) Optimal time transfer in bus transit route network design using a genetic algorithm, *Journal of Transportation Engineering*, 129 (5), 510–521.

Olivier, J.G.J. *et al.*, (2005) Recent trends in global greenhouse gas emissions: regional trends 1970-2000 and spatial distribution of key sources in 2000. *Environmental Science*, 2(2-3), pp. 81-99.

Olivier, J.G.J., *et al.*, (2006) Part III: Greenhouse gas emissions: 1. Shares and trends in greenhouse gas emissions; 2. Sources and Methods; Greenhouse gas emissions for 1990, 1995 and 2000. In CO_2 emissions from fuel combustion 1971-2004, 2006 Edition, pp. III.1-III.41. International Energy Agency (IEA), Paris

Palmer, A., (2004) The environmental implications of grocery home delivery, ELA doctorate workshop, Centre for Logistics and Supply Chain Management, Cranfield University

Pardalos, P.M.(Ed.)., (2002) Combinatorial and Global Optimization. River Edge, NJ, USA: World Scientific. p 10pq.

Patil, G. R., & Ukkusuri, S. V. (2007). System-optimal stochastic transportation network design. *Transportation Research Record*, (2029), 80-86. doi: 10.3141/2029-09

Pattnaik, S.B., Mohan, S., Tom, V.M., (1998) Urban bus transit network design using genetic algorithm, *Journal of Transportation Engineering*, 124 (4), 368–375.

Poorzahedy, H., Abulghasemi, F., (2005) Application of ant system to network design problem, *Transportation*, 32 (3), 251–273.

Poorzahedy, H., Rouhani, O.M., (2007) Hybrid meta-heuristic algorithms for solving network design problem, *European Journal of Operational Research*, 182 (2), 578–596.

Poorzahedy, H., Turnquist, M.A., (1982) Approximate algorithms for the discrete network design problem, *Transportation Research Part B*, 16 (1), 45–55.

Pronello C., André M., (2000) Pollutant emissions estimation in road transport models. INRETS-LTE report, vol 2007

Qin, J., Ni, L. L., & Shi, F. (2013). Mixed Transportation Network Design under a Sustainable Development Perspective. *Scientific World Journal*. doi: 10.1155/2013/549735

Reeves, Colin R., (2002) Genetic Algorithms - Principles and Perspectives : A Guide to GA Theory. Secaucus, NJ, USA: Kluwer Academic Publishers, p1.

Rushton, Alan, (2006) Handbook of Logistics and Distribution Management (3rd Edition). London, GBR: Kogan Page, Limited, p 579.

Russo, F., Vitetta, A., (2006) A topological method to choose optimal solutions after solving the multi-criteria urban network design problem, *Transportation*, 33, 347–370.

Sarker R., (2008) Optimization Modelling: a practical approach. CRC Press

Sarker, R. (Ed.), (2002) Evolutionary Optimization. Secaucus, NJ, USA: Kluwer Academic Publishers, p 29.

Sbihi, A., Eglese, R.W., (2007) Combinatorial optimization and Green logistics. *4OR: A Quarterly Journal of Operations Research*, 5 (2), 99–116.

Sergienko, I.V., Hulianytskyi, L.F., Sirenko, S.I., (2009) Classification of applied methods of combinatorial optimization. *Cybernetics and Systems Analysis*, 45 (5), 732-741

Sharma, S., Mathew, T. V., & Ukkusuri, S. V. (2011). Approximation Techniques for Transportation Network Design Problem under Demand Uncertainty. *Journal of Computing in Civil Engineering*, 25(4), 316-329. doi: 10.1061/(asce)cp.1943-5487.0000091

Sharma, S., Ukkusuri, S. V., & Mathew, T. V. (2009). Pareto Optimal Multiobjective Optimization for Robust Transportation Network Design Problem. *Transportation Research Record*, (2090), 95-104. doi: 10.3141/2090-11

Sivanandam S.N. and Deepa S.N., (2008) Introduction to Genetic Algorithms, Terminologies and Operators of GA, Springer, pp 39-81

Solanki, R.S., Gorti, J.K., Southworth, F. (1998) The highway network design problem, *Transportation Research Part B*, 32, 127–140.

Stock, James R.(Ed.), (2002) Qualitative methods and approaches in logistics: part 2. Bradford, GBR: Emerald Group Publishing Limited, p 24.

Suwansirikul, C., Friesz, T.L., Tobin, V. (1987) Equilibrium decomposed optimization: A heuristic for the continuous equilibrium network design problem, *Transportation Science*, 21, pp. 254–263.

Ukkusuri, S. V., & Patil, G. (2009). Multi-period transportation network design under demand uncertainty. *Transportation Research Part B-Methodological*, 43(6), 625-642. doi: 10.1016/j.trb.2009.01.004

Ukkusuri, S. V., Mathew, T. V., & Waller, S. T. (2007). Robust transportation network design under demand uncertainty. *Computer-Aided Civil and Infrastructure Engineering*, 22(1), 6-18. doi: 10.1111/j.1467-8667.2006.00465.x

Ukkusuri, S.V., Mathew, T.V., Waller, S.T., (2007) Robust transportation network design under demand uncertainty, *Computer-Aided Civil and Infrastructure Engineering*, 22 (1), 6–18.

Unnikrishnan, A., & Waller, S. T. (2009). Freight Transportation Network Design Problem for Maximizing Throughput Under Uncertainty. *Transportation Research Record*, (2090), 105-114. doi: 10.3141/2090-12

Wang, H., Xiao, G. Y., Zhang, L. Y., & Ji, Y. B. B. (2014). Transportation Network Design considering Morning and Evening Peak-Hour Demands. *Mathematical Problems in Engineering*. doi: 10.1155/2014/806916

WBCSD, (2002) Mobility 2001: World Mobility at the End of the Twentieth Century, and its Sustainability. World Business Council for Sustainable Development

WBCSD, (2004a) Mobility 2030: Meeting the Challenges to Sustainability. <http://www.wbcsd.ch/> accessed 30/05/07.

WBSCD, (2004b) IEA/SMP Model Documentation and Reference

Projection. Fulton, L. and G. Eads, <http://www.wbcsd.org/web/publications/mobility/smp-model-document.pdf> accessed 30/05/07

Xu, J. P., Gang, J., & Lei, X. (2013). Hazmats Transportation Network Design Model with Emergency Response under Complex Fuzzy Environment. *Mathematical Problems in Engineering.* doi: 10.1155/2013/517372

Yildiz, T., Yercan, F., (2010) The cross-entropy method for combinatorial optimization problems of seaport logistics terminal. *Journal Transport*, 25 (4), 411-422.

Zhang, H. Z., & Gao, Z. Y. (2009). Bilevel programming model and solution method for mixed transportation network design problem. *Journal of Systems Science & Complexity*, 22(3), 446-459. doi: 10.1007/s11424-009-9177-3

ABOUT THE AUTHOR

Turkay Yildiz received his Ph.D. from the Institute of Marine Sciences and Technology, Dokuz Eylul University, Izmir, Turkey. He received his Master's Degree in Logistics Management from Izmir University of Economics. He has a number of peer reviewed publications and conference presentations at various countries in such fields as transportation, logistics and supply chains. He also has various levels of expertise in the applications of Information Technology.

www.ingramcontent.com/pod-product-compliance
Lightning Source LLC
Chambersburg PA
CBHW070609300426
44113CB00010B/1466